Mathematical Puzzles and Curiosities

Problem Solving in Mathematics and Beyond

Print ISSN: 2591-7234
Online ISSN: 2591-7242

Series Editor: Dr. Alfred S. Posamentier
Distinguished Lecturer
New York City College of Technology - City University of New York

---

There are countless applications that would be considered problem solving in mathematics and beyond. One could even argue that most of mathematics in one way or another involves solving problems. However, this series is intended to be of interest to the general audience with the sole purpose of demonstrating the power and beauty of mathematics through clever problem-solving experiences.

Each of the books will be aimed at the general audience, which implies that the writing level will be such that it will not be engulfed in technical language — rather the language will be simple everyday language so that the focus can remain on the content and not be distracted by unnecessarily sophiscated language. Again, the primary purpose of this series is to approach the topic of mathematics problem-solving in a most appealing and attractive way in order to win more of the general public to appreciate this most important subject rather than to fear it. At the same time we expect that professionals in the scientific community will also find these books attractive, as they will provide many entertaining surprises for the unsuspecting reader.

Published

Vol. 43 *Mathematical Puzzles and Curiosities*
by Ivo Fagundes David de Oliveira, Tanya Khovanova and Yogev Shpilman

Vol. 42 *Learning Geometry Through Problem Solving*
by Boris Pritsker

Vol. 41 *Maxima and Minima: How Extremes Can Explain Mathematics Concepts and Facilitate Problem Solving*
by Hans Humenberger and Alfred S Posamentier

Vol. 40 *Appreciating Geometry Through Extraordinary Relationships*
by Alfred S Posamentier

Vol. 39 *Fun Math: Problem Solving Beyond the Classroom*
by Alfred S Posamentier

For the complete list of volumes in this series, please visit www.worldscientific.com/series/psmb

Problem Solving in Mathematics and Beyond
Volume 43

Mathematical Puzzles and Curiosities

Ivo Fagundes David de Oliveira
Tanya Khovanova
Yogev Shpilman

World Scientific

NEW JERSEY • LONDON • SINGAPORE • BEIJING • SHANGHAI • HONG KONG • TAIPEI • CHENNAI • TOKYO

Published by

World Scientific Publishing Co. Pte. Ltd.
5 Toh Tuck Link, Singapore 596224
USA office: 27 Warren Street, Suite 401-402, Hackensack, NJ 07601
UK office: 57 Shelton Street, Covent Garden, London WC2H 9HE

Library of Congress Control Number: 2025042226

British Library Cataloguing-in-Publication Data
A catalogue record for this book is available from the British Library.

Problem Solving in Mathematics and Beyond — Vol. 43
MATHEMATICAL PUZZLES AND CURIOSITIES

ISBN 978-981-98-2341-3 (hardcover)
ISBN 978-981-98-2236-2 (paperback
ISBN 978-981-98-2237-9 (ebook for institutions)
ISBN 978-981-98-2238-6 (ebook for individuals)

For any available supplementary material, please visit
https://www.worldscientific.com/worldscibooks/10.1142/14561#t=suppl

Desk Editors: Nambirajan Karuppiah/Srinidhi Murugan

Typeset by Stallion Press
Email: enquiries@stallionpress.com

Contents

Chapter 1

This Title Has Twenty ________ Letters
?

Welcome to the whimsical world of self-referential puzzles. In this chapter, we invite you to step into a peculiar shop, where every item carries a self-referential twist, producing unconventional paradoxical riddles.

Imagine walking into this extraordinary shop, where shelves are lined with vouchers, tickets, and coupons that reference themselves in the most curious ways. As you cross the threshold, you are greeted by the shop owner, who begins with a perplexing introduction: "Before I begin speaking, I'd like to say a few words...". And with that, he launches into a convoluted yet fascinating monologue about the latest self-referential items you can purchase...

As you browse through the shop, you'll encounter tickets that declare their own falsity, coupons that offer discounts on themselves, and vouchers that question their own validity. Each item you pick up poses a unique puzzle, an enigma wrapped in a riddle, waiting for you to unravel its secrets. Are you ready to explore the paradoxical pleasures of the self-referential shop? Step inside and let the puzzles begin.

Puzzle 1.1 **Low Budget**

As you walk into the self-referential shop, you see a shelf full of vouchers. A voucher reading "The next voucher you buy costs N times its price tag" has a price tag of N dollars. Each voucher's multiplier applies only to the immediately following purchase. An infinite number of vouchers are in front of you, corresponding to $N = 1, 2, 3, \ldots$ There is a single voucher available to purchase for each value of N. If, for example, you bought a voucher corresponding to $N = 2$, followed by one corresponding to $N = 3$, followed by one corresponding to $N = 4$, you'd pay \$20 in total (\$2 for the first voucher, plus $\$3 \cdot 2 = \$6$ for the second one, plus $\$4 \cdot 3 = \$12$ for the third one).

Question. *Given a budget of 26 dollars, what is the maximum number of vouchers you can buy?*

Puzzle 1.2 **Truthful Coupons**

The self-referential shop also sells special coupons. A coupon reading "The number of false coupons you bought is a positive integer multiple of N" costs N dollars. An infinite number of coupons are in front of you, corresponding to $N = 1, 2, 3, \ldots$ There is an unlimited number of coupons available to purchase for each value of N. After the purchase, you are allowed to label each coupon true or false as long as the labeling is consistent.

For example, if you bought three copies of the 2-dollar coupon and four copies of the 3-dollar coupon, you can label the 2-dollar coupons as true and the 3-dollar coupons as false. You can also reverse the labels. In the first labeling, you have three true coupons, and in the second one you have four. However, labeling all seven coupons true (or false) is inconsistent and not allowed.

Question. *Given a budget of 31 dollars to buy coupons, what is the maximum number of true coupons in your collection?*

> **Curiosity.** **Autobiographical Numbers**
> A number is called *autobiographical* if the first digit describes the number of zeros in this number, the second digit describes the number of ones, and so on. The smallest autobiographical number is 1210 as it contains one 0-digit, two 1-digits, one 2-digit, and zero 3-digits. Can you find the only 10-digit autobiographical number?

Puzzle 1.3 — One of a Kind

Some of the high-end products in the self-referential shop are the self-referential tickets. A ticket reading "The number of dollars you spent today on tickets is a positive integer multiple of N" costs N dollars. An infinite number of tickets are in front of you, corresponding to $N = 1, 2, 3, \ldots$ There is a single ticket available to purchase for each value of N. After your purchases, a ticket is true precisely when the total amount you spent on tickets is an integer multiple of its own N; otherwise, it is false.

Question. *Given a budget of 31 dollars, what is the maximum number of tickets you can buy today so that all of them would be true?*

Puzzle 1.4 — I'll Take One of Each

The self-referential shop sells membership cards that give discounts on subsequent purchases. A membership card that costs N dollars reduces the cost of each subsequent purchase to $\frac{1}{N}$ times its original price. The membership's discount also applies if you want to buy more membership cards. If several cards are owned, their discounts apply multiplicatively (so owning cards N and M multiplies the price of further purchases by $\frac{1}{NM}$). Today, while you visit the shop, there is one copy of each card with $N = 1, 2, 3, \ldots$ in stock.

Question. *What is the cheapest order to purchase the first ten cards?*

Curiosity. **Pinocchio**

The image above depicts Pinocchio sitting in the book alley of the self-referential shop, wondering what would happen to his nose if he said “My nose will grow now”. This is known as Pinocchio’s Paradox.

Puzzle 1.5 **Pirate Auction**

On special occasions, the self-referential shop holds unusual auctions. Today, there is a pirate treasure map for sale through one such auction. The current owner, a pirate, says: "This treasure map is for sale! But you can buy it only if you are able to guess its price correctly. However, after each incorrect guess, I will add to the price a fixed predetermined amount". For example, if the initial price is \$10 and the fixed increase after each incorrect guess is \$1, to obtain the right to purchase it, you would have to guess \$10 for the first guess, or \$11 for the second guess, or \$12 for the third guess, and so on. You know neither the initial price of the map nor the fixed increment amount (but both are positive integers), and after each guess, the pirate replies only by "right" or "wrong".

Question. *Can you find a way to guarantee correctly guessing the price after a finite number of guesses?*

Puzzle 1.6 **IQ Tests**

While browsing the self-referential shop, you see the following poster:

If you choose the answer to this question at random, what is the probability you will be correct?

(a) 25% **(b) 50%** **(c) 60%** **(d) 25%**

Buy an instant multiple-choice IQ test here.

Multiple-choice IQ tests are printed instantly for your convenience at the shop. The test asks the same question as the poster and lists several answer options, each an integer from 1 to 100, followed by a percent sign %; options may repeat. To buy an IQ test, you must select a cost C and a number of options $K > 0$. The text printed out will have the following property:

There will be exactly K numbers printed on the customized test, and they will add up to the cost C selected.

It follows that your K can't exceed C. The example above has $K = 4$ options with a cost of $C = 160$. Answers 50 and 60 are chosen with probability 25%, while answer 25 is chosen with probability 50%. Thus, our sample test doesn't have a correct answer. Here you can find three more examples of what other tests might look like.

	Cost	# of Options	Sample Test Printed Out
Test 1	**$100**	5	(a) 25% (b) 25% (c) 20% (d) 20% (e) 10%
Test 2	**$50**	4	(a) 25% (b) 10% (c) 10% (d) 5%
Test 3	**$250**	4	(a) 75% (b) 75% (c) 75% (d) 25%

Most tests don't seem to have a correct answer available, but if you are lucky, they sometimes do. In the examples above, only test 2 has a correct answer. Let's understand why.

Test 1. In this test, the option 10% is chosen 20% of the time while options 20% and 25% are chosen 40% of the time. Thus, this test also has no correct answer.

Test 2. This test has the correct answer of 25%.

Test 3. This case is a little bit trickier since the number 75% shows up on 75% of the answers, making it seem like 75% is the correct answer. However, with a similar argument, you will notice that 25% also seems to be correct! Therefore, 100% of the solutions seem to be correct, and thus, if there were a correct answer, the correct answer should have been 100% and not 25% nor 75%.

Question (1). *Suppose you wish to purchase a $100 multiple-choice IQ test with a correct answer. What is the maximum number of options the test may have? What is the correct answer for that test?*

Question (2). *What is the price of the cheapest IQ test with a correct answer?*

Curiosity. **A Self-Referential Gem**

This is called *Tupper's self-referential formula*:

$$\frac{1}{2} < \left\lfloor \text{mod}\left(\left\lfloor \frac{y}{17} \right\rfloor 2^{-17\lfloor x \rfloor - \text{mod}(\lfloor y \rfloor, 17)}, 2 \right) \right\rfloor.$$

"What can be so special about this strange inequality?", we hear you ask. Well, imagine you were to color the x-plane, filling every dot in black if it satisfies the inequality. If you search across the plane, you will find the following image[a]:

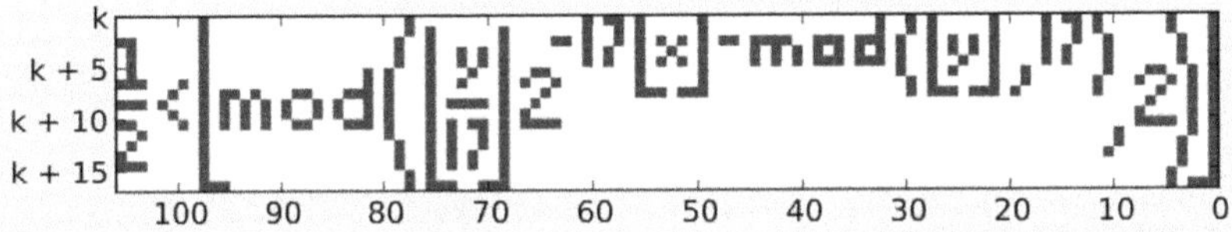

[a] Attention! In Tupper's convention, the x- and y-axis are mirrored, and the constant k in the y-axis is a 543-digit integer!

This page was unintentionally left blank

Solutions

It is a well-known secret at the heart of mathematics that self-referencing should never be done carelessly. One beautiful example of this comes from the famous logician Bertrand Russell when considering the simple question of how to define a *set*. You see, normally, a set is just a collection of objects, like "all chairs with three legs" or "all prime numbers", but what happens if we allow for self-referencing when we define a set? Bertrand Russell pulls our rug with a simple story about a barber who swore an oath to shave all those, and only those, who don't shave themselves. Consider it carefully: who do you think should shave the barber?

If he shaves himself, he breaks his oath by shaving someone who shaves himself! And if he doesn't shave himself, then yet again he breaks his oath by not shaving *all* those who do not shave themselves!

This small example shows how careless self-referencing might lead to complex and seemingly unsolvable paradoxes at the heart of mathematics, that is, logic itself!

As dangerous as self-referencing might be, we chose this opening chapter to be about self-referential puzzles for a reason. And, you can be sure that throughout this book, this will not be the last you will see of self-referential twists and paradoxes!

Solution to Puzzle 1.1 — Low Budget

To maximize the number of vouchers we can buy, it makes intuitive sense to purchase the most expensive one last, i.e., at the end of the shopping spree. This way, its high price tag will not multiply the price of any other voucher. It also makes sense to buy the cheapest voucher, corresponding to $N = 1$, right before the most expensive one. That way, the high price of the expensive voucher will not be multiplied by any number greater than 1. Using these tricks and by simple elimination, we find that you can buy five vouchers, which can be done in two different ways: in the order 32415 or the order 42315. Either way, the cost would be $3+6+8+4+5 = 4+8+6+3+5 = 26$.

If you wish to verify that the above tricks lead to a correct answer, one way is to verify all possible orders of the five cheapest vouchers: that makes for $5! = 120$ possibilities. Instead of doing that, we will verify by deduction that these are the best ways to buy five vouchers. Let us indicate the printed value of the ith voucher we buy as V_i. This way, the cost of purchasing five vouchers in a given order is:

$$\text{Total Cost} = V_1 + V_1V_2 + V_2V_3 + V_3V_4 + V_4V_5.$$

First, let us recognize that we must buy the five cheapest vouchers. This is easy to see since if any other higher value voucher were purchased, the Total Cost above would be reduced by replacing that voucher with the missing cheaper voucher. Second, let us assume that V_5 is not equal to 5 to see that that is not an optimal choice: If instead V_4 were equal to 5, then switching the values of V_4 and V_5 would reduce the Total Cost. A similar argument (a little bit more detailed, though) also shows that V_1, V_2, and V_3 could not be equal to 5 since another set of swaps would reduce the cost, thus guaranteeing that $V_5 = 5$ gives a reduced price. Therefore, we have:

$$\text{Total Cost} = V_1 + V_1V_2 + V_2V_3 + V_3V_4 + 5V_4.$$

Another set of replacement arguments can show that V_4 must equal 1. Thus, leaving us with the total cost of

$$\text{Total Cost} = V_1 + V_1V_2 + V_2V_3 + V_3 + 5,$$

which is totally symmetric between V_1 and V_3. Now, all we need to verify is that the larger terms should be V_1 and V_3, and the smallest term should be V_2. Thus, giving us both sequences: 32415 and 42315. Using similar arguments, one can check that there is not enough money for six vouchers.

One surprisingly clever trick:

Tanya gave this puzzle to her students, and Soham came up with the following proof that the number of vouchers is fewer than six. Obviously, if we buy six vouchers in an optimal way, we buy vouchers 1 to 6 in some order. The corresponding cost is

$$V_1 + V_1V_2 + V_2V_3 + V_3V_4 + V_4V_5 + V_5V_6,$$

which is greater than

$$V_1V_2 + V_3V_4 + V_5V_6.$$

By using the AM–GM inequality (the arithmetic mean is not less than the geometric mean: $\frac{x+y+z}{3} \geq \sqrt[3]{xyz}$), we get that the total cost is at least

$$3\sqrt[3]{V_1V_2V_3V_4V_5V_6} \geq 3\sqrt[3]{720} > 26.$$

Source: This puzzle was developed by the authors in 2024.

Solution to Puzzle 1.2 Truthful Coupons

What is curious about this puzzle is that if any coupon is labeled false, then for sure, the coupon "The number of false coupons you bought is a positive integer multiple of 1" must be true. These are also the cheapest coupons, and thus, you might be tempted to spend all your money on them. However, if you buy 31 coupons saying, "The number of false coupons you bought is a positive integer multiple of 1," then labeling all of them true creates a contradiction! Therefore, you are forced to buy at least one false coupon: "The number of false coupons is a positive integer multiple of 2," giving the solution: a maximum of 29 true coupons.

Source: This puzzle was developed by the authors in 2024.

Curiosity. **Quines**
A *quine* is a computer program that takes no input and prints a copy of its own source code. Challenge: Can you write a query to a chatbot so that it will reply with exactly what you wrote?

Solution to Puzzle 1.3 **One of a Kind**

The sum of the tickets' prices that were purchased should be divisible by the price of each ticket. When a number equals the sum of all its proper divisors, it is called a *perfect number*. Thus, one strategy is to make the sum of the tickets equal to the largest perfect number that is smaller than 31, which is 28. This strategy would give you all the tickets corresponding to the divisors of 28 (excluding itself): $1+2+4+7+14=28$. A total of five tickets, but do perfect numbers give us the perfect strategy? Let's check! From 1 to 31, here are all the numbers that have six or more divisors:

Number	Divisors	Number of Divisors
12	1, 2, 3, 4, 6, 12	6
18	1, 2, 3, 6, 9, 18	6
20	1, 2, 4, 5, 10, 20	6
24	1, 2, 3, 4, 6, 8, 12, 24	8
28	1, 2, 4, 7, 14, 28	6
30	1, 2, 3, 5, 6, 10, 15, 30	8

If there is a solution with six or more tickets, then the numbers $12, 18, 20$, and 28 can't be it because the sum of their six divisors (including themselves) either exceeds the $31 budget or the sum is not divisible by all the terms. Therefore, we only need to check 24 and 30 if they have a subset of divisors larger than or equal to 6 that adds up to 24 or 30, respectively.

For the number 30, it is not possible to find such a subset of divisors, but for the number 24, we find that 24 equals $1+2+3+4+$

6 + 8. Therefore, it is possible to buy six true tickets for 24 dollars (a less-than-perfect number), i.e., the answer is 6.

Curiosity. **Semiperfect Numbers**
A number that is equal to the sum of all or some of its proper divisors is called a semiperfect number.

Bonus. Summing up the inverses of the divisors of 28, a perfect number, gives us

$$\frac{1}{1}+\frac{1}{2}+\frac{1}{4}+\frac{1}{7}+\frac{1}{14}+\frac{1}{28}=2.$$

Is it a coincidence that the sum of these fractions is a whole number? What happens when you add up the inverses of the divisors of other perfect numbers?

Curiosity. **Perfect Numbers**
The four smallest perfect numbers are 6, 28, 496, and 8128. Euclid studied perfect numbers back in 300 BC! He was able to prove that if $2^p - 1$ is a prime number, then $(2^p - 1)2^{p-1}$ is a perfect number (the smallest perfect numbers correspond to inserting $p = 2, 3, 5, 7$ in this formula). Over 2000 years after Euclid, Euler showed that this formula yields all even perfect numbers. Up to the date of writing this book, it is unknown whether there is an infinite number of perfect numbers or whether an odd perfect number exists. Solving one of these puzzles will surely bring you fame and glory.

Source: This puzzle was developed by the authors in 2024.

Solution to Puzzle 1.4 — I'll Take One of Each

The answer is to buy the membership cards in the order 2, 3, 4, 5, 6, 7, 8, 9, 10, 1. Here is the proof.

First, let us recognize that 1 must be the last card to be purchased. To see this, notice that it doesn't give any discounts, and therefore, it is always beneficial to postpone this card's purchase as much as possible.

Now, let's look at two tickets in the middle (or the beginning) of the sequence of purchases: call them tickets a and b. Since they are somewhere in the middle, we know that both a and b are greater than 1. Let's compare the cost associated with the sequence $\dots ab \dots$ and $\dots ba \dots$ with all terms other than a and b being equal. The cost for all the tickets that are not a or b is the same for both sequences. Thus, the sequence $\dots ba \dots$ is cheaper if and only if:

$$\frac{a}{z} + \frac{b}{az} < \frac{b}{z} + \frac{a}{bz},$$

where z is the discount provided by all purchases preceding a and b. The above inequality is equivalent to

$$a^2 b + b^2 < b^2 a + a^2,$$

which in turn is equivalent to

$$a^2(b-1) < b^2(a-1).$$

As both a and b are greater than 1, we can divide both sides of the inequality above by $(a-1)(b-1)$, getting that

$$\frac{a^2}{a-1} < \frac{b^2}{b-1}.$$

Since $\frac{x^2}{x-1}$ is an increasing function for $x > 1$, we conclude that the cost of $\dots ab \dots$ is less than that of $\dots ba \dots$ if and only if $a < b$.

From this, we conclude that all terms (except for the last) must be in increasing order to minimize the cost. Therefore, the solution must be 2, 3, 4, 5, 6, 7, 8, 9, 10, 1. When rounding to cents, this adds up to \$4.44.

Curiosity. **An Infinite Discount**

Can you guess how much money you would need in order to purchase one copy of each membership card from 1 to ∞?

PS: If we leave card 1 for last (like we did in the above puzzle solution), do we ever buy it at all?

Answer: *You would need $2e - 1$ dollars. This is approximately \$4.43656 which, when rounding to cents, is again \$4.44. The same price you would pay for the first 7 cards alone.*

Source: This puzzle was developed by the authors in 2025.

Solution to Puzzle 1.5 **Pirate Auction**

At first glance, it might seem impossible to guess the price of the map correctly. If the price remained constant between each guess, you could start guessing with \$1, \$2, \$3, ... and eventually reach the correct price. But when the price increases between guesses, you might never reach it at the correct moment with this strategy. So, how is it possible to guarantee a correct guess after a finite number of attempts?

Let us label the initial price of the map by a_0, and the fixed amount by which the price increases after every inaccurate guess by d. Then, the treasure's cost after n inaccurate guesses, a_n, is $a_n = a_0 + nd$. If we keep track of the number of failed attempts, n, for every guess, we can calculate the hypothetical price of the treasure for a pair a_0 and d.

Now, we enumerate all possible pairs of a_0 and d, both being natural numbers. When you want to scan over all possible values of a_0 and d, you might want to fix a_0, then scan over all possible values of d, and then increase the value of a_0 by one and repeat. However, since there are infinite possible values of d (and of a_0), this is not going to work. The proper way of doing this, instead of fixing a_0 or d while scanning the other, is to fix the sum $a_0 + d$ while scanning over all values of a_0 and d that produce that sum. Then, you increase the sum by 1 and repeat.

In the following diagram, we provide a visualization of this process.

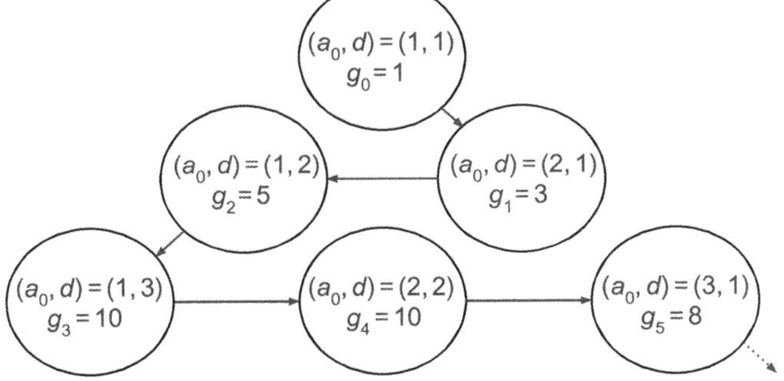

Each row has a constant sum of a_0+d and goes through all options of a_0 and d that produce that sum. Since a_0 and d are positive integers, their smallest sum could be 2, corresponding to $a_0 = 1$ and $d = 1$. Then, our initial guess g_0 (with zero inaccurate guesses) would be: $g_0 = a_0 = 1$. If that was incorrect, we now increase the sum by 1 and consider all values of a_0 and d that give $a_0 + d = 3$. There are two options for that: $a_0 = 1$, $d = 2$ and $a_0 = 2$, $d = 1$. For guess g_1, following the order in the diagram above, we have $a_0 = 2$, $d = 1$; with one incorrect guess, this corresponds to a cost of 3, so our next guess is $g_1 = a_1 = 3$. Again, following the order in the diagram above, guess g_2 corresponds to $a_0 = 1$, $d = 2$; and, with two incorrect guesses, to catch this pattern we need to guess $g_2 = a_2 = 5$. This method was used in "Hilbert's Paradox of the Grand Hotel" [3] to show that pairs of integers can be enumerated by non-negative integers. It is called "Cantor pairing function".

Source: Folklore.

Solution to Puzzle 1.6 **IQ Tests**

Solution to question 1:
We paid \$100 for the test, so the sum of values of all options is 100 as well. The maximum number of options that a correct test can have is 80, with 5% being the correct answer. Here is one example of a test with that many options for \$100:

$$[\ 5\%\ 5\%\ 5\%\ 5\%\ 3\%\ 3\%\ 1\%\ \ldots\ 1\%]$$

(this example has 74 copies of the number 1%)

One can verify that 5% is the correct answer since it appears exactly in four out of 80 slots, i.e., 5% of the slots, and no other number has this property.

Now, to prove that the solution above has, in fact, the biggest number of options, first notice that the longest solvable IQ test must have fewer than 100 options since the only \$100 IQ test with 100 options has 100 copies of 1%. Therefore, we only need to verify the values of K between 81 and 99.

Since the percentages on the IQ tests are all integer values, if the IQ test has K options and the correct answer is N, then the correct answer must appear in exactly $\frac{NK}{100}$ options.

Therefore, we only need to check values of N and K such that NK is a multiple of 100. For example, if a solvable IQ test had $K = 82$ slots, then the minimum possible solution would have to be $N = 50$ because $N = 50$ is the minimum number such that NK makes a multiple of 100. With the same reasoning, we see that for K equaling 86, 94, and 98, the answer N has to be at least 50. Similarly, for K equaling 81, 83, 87, 89, 91, 93, 97, and 99, the answer N has to be at least 100. And, continuing this pattern, we can verify other values of K between 81 and 99. See the following table for details.

$K = 81 \Rightarrow N \geq 100$	$K = 86 \Rightarrow N \geq 50$	$K = 91 \Rightarrow N \geq 100$	$K = 96 \Rightarrow N \geq 25$
$K = 82 \Rightarrow N \geq 50$	$K = 87 \Rightarrow N \geq 100$	$K = 92 \Rightarrow N \geq 25$	$K = 97 \Rightarrow N \geq 100$
$K = 83 \Rightarrow N \geq 100$	$K = 88 \Rightarrow N \geq 25$	$K = 93 \Rightarrow N \geq 100$	$K = 98 \Rightarrow N \geq 50$
$K = 84 \Rightarrow N \geq 25$	$K = 89 \Rightarrow N \geq 100$	$K = 94 \Rightarrow N \geq 50$	$K = 99 \Rightarrow N \geq 100$
$K = 85 \Rightarrow N \geq 20$	$K = 90 \Rightarrow N \geq 10$	$K = 95 \Rightarrow N \geq 20$	

Any IQ test with the above numbers will burst the budget of $100. Thus, the maximum number of options is 80. We gave one example of such a test with the correct answer 5%. Are there any other examples with a different correct answer? As we explained before, the correct answer N has to be a multiple of 5. Suppose it is $5m$, then it has to appear in $\frac{5m \cdot 80}{100} = 4m$ options. We can check again that if $m > 1$, we go over budget. If $N = 5$, we can have several tests, as we can distribute 80 leftover percentage points into 76 leftover slots in several different ways.

Solution to question 2:

Assume for now that the IQ test has the following format: [1% 1% ... 1% X%], where $X > 1$. We will first show (in *Part 1*) that the cheapest solvable IQ test with this format repeats the number 1% exactly nine times and that $X = 10\%$, and therefore the IQ Test costs $19. We will then show (in *Part 2*) that this is indeed the minimum cost of a solvable IQ test.

Proof Part 1. Let's assume $1 repeats M times. The probability to choose X is $\frac{1}{M+1}$, which should equal $\frac{X}{100}$. Then, $X = \frac{100}{M+1}$. The price of such a test is $M + X = M + \frac{100}{M+1}$, which is minimized for $M = 9$.

Proof Part 2. Assume there exists a solvable IQ test that cost \$18 or less. This implies that there must be 18 or fewer options on the IQ test. Let K be the number of options and N be the correct answer on the IQ test. As mentioned in the solution to Question 1, the correct answer must appear in exactly $\frac{NK}{100}$ slots, and therefore, NK must be a multiple of 100. With this, we can deduce that if there were $K = 18$ options, the correct answer N must be greater than or equal to 50, which would already exceed the cost of \$19 we found in part 1. Following this reasoning, we verify in the table below what happens for lower values of K.

$K = 1 \Rightarrow N \geq 100$	$K = 5 \Rightarrow N \geq 20$	$K = 9 \Rightarrow N \geq 100$	$K = 13 \Rightarrow N \geq 100$
$K = 2 \Rightarrow N \geq 50$	$K = 6 \Rightarrow N \geq 50$	$K = 10 \Rightarrow N \geq 10$	$K = 14 \Rightarrow N \geq 50$
$K = 3 \Rightarrow N \geq 100$	$K = 7 \Rightarrow N \geq 100$	$K = 11 \Rightarrow N \geq 100$	$K = 15 \Rightarrow N \geq 20$
$K = 4 \Rightarrow N \geq 25$	$K = 8 \Rightarrow N \geq 25$	$K = 12 \Rightarrow N \geq 25$	$K = 16 \Rightarrow N \geq 25$
			$K = 17 \Rightarrow N \geq 100$

Therefore, by inspection, we verify that the number of options K must equal 10 if we wish to stay below the budget of \$19. It is easy to verify that for $K = 10$, the only possible values for N are 10, 20,…, 90, 100; and for $N = 10$ it is clear that [1% 1%,…, 1% 10%] both minimizes the cost and stays below \$20 (which is a lower bound on the cost of any other possible value of N). This completes the proof.

The answer to this puzzle, \$19, feels a bit arbitrary, doesn't it? Well, the ancient game of Go is customarily laid out on a 19×19 grid, which is also somewhat peculiar.

Source: This puzzle was inspired by a viral post that has been circulating on multiple social media platforms. The oldest reference we found was posted in 2010 on Reddit by a deleted user: https://www.reddit.com/r/pics/comments/c64gp/multiple_choice_if_you_choose_an_answer_to_this/.

> *Wanting to know what he looked like while he was sleeping, an idiot stood in front of a mirror with his eyes closed.*
> A joke from the 4th century, found in *Philogelos*, the oldest joke book in history.

Chapter 2

Knights, Nuts & Knaves

Welcome to the precise world of logic puzzles. Traditionally, such puzzles take place on an island, where there are two kinds of people: *knights*, who always tell the truth, and *knaves*, who always lie. In this chapter, we invite you to step onto a peculiar island where, in addition to the standard two categories, we have people of a third kind who are just a little bit... different. We call them *nuts*. Nuts have a quirky habit of strictly alternating between truth and lies.

For instance, if you hear an islander say, "Today is Wednesday. Today is Tuesday," you'd know that the islander is not a knight, as at least one of the statements must be false. But what is this person? You can figure it out if you know what day it is. If today is neither Tuesday nor Wednesday, the person must be a knave. Otherwise, they must be a nut because that's just how they roll!

You will meet a lot of interesting characters, all of them are from this island, including the good detective Radstein, who investigates crimes. It doesn't matter what type of islander our detective is, but you can assume that he is a knight. In the first four puzzles, you will help Detective Radstein crack curious crimes. In the remaining puzzles, you will wander around this beautiful island and learn more about its inhabitants.

Remember the three rules of this chapter!

- Knights always tell the truth.
- Knaves always lie.
- Nuts alternate between the two.

Puzzle 2.1 **Three Nuts**

Detective Radstein is investigating an arson. He apprehends three suspects: Anne, Bill, and Caroline. The detective knows that no one else could have participated in the crime. During the interrogation, the suspects made the following statements:

- Anne: I didn't do it. Bill did it alone.
- Bill: — remains silent —
- Caroline: I didn't do it. Bill didn't do it.

Detective Radstein also discovered that all three suspects were nuts.

Question. *Who committed the arson?*

Curiosity. **I Remain Silent**
In the US, you have the right to remain silent if legal authorities question you. However, if you want to exercise this right, you must say it out loud.

Puzzle 2.2 **Can You Repeat That?**

Detective Radstein is investigating a robbery. He apprehends two suspects: Avigail and Benjamin. The detective knows that one of them participated in the robbery. During the interrogation, the suspects made the following statements:

- Avigail: I don't know which of us is a nut. I didn't commit the robbery. Today is Tuesday.
- Benjamin: I don't know which of us is a nut. I didn't commit the robbery. Today is Tuesday indeed.

The detective knows that in this particular village, nuts know each other, while others don't know who the other people are. After the suspects' statements, he immediately arrested one of them.

Question. *Who committed the crime?*

Curiosity. **I Am a Knave**

Have you heard this famous puzzle about knights and knaves? You are touring the knights-and-knaves island when an islander approaches you and says, "I am a knave." Who is the islander?

This puzzle is a paradox, as neither a knight nor a knave can say that sentence. However, on our island, this puzzle is no longer a paradox. Can you solve it?

Puzzle 2.3 **Together or Not?**

Detective Radstein is investigating a burglary. He apprehends three suspects: Allan, Bruno, and Caio. The detective knows that no one else could have participated in the burglary. During the interrogation, the suspects made the following statements:

- Allan: I didn't do it. Bruno and Caio both participated in the burglary.
- Bruno: I didn't do it. Caio participated in the burglary.
- Caio: — remains silent —

Detective Radstein also discovered that all three suspects were nuts.

Question. *Who committed the burglary?*

Puzzle 2.4 **I Did It**

Detective Radstein is investigating a murder. He apprehends three suspects: Anne, Bill, and Caroline. The detective knows that exactly one of them committed the murder. During the interrogation, the suspects made the following statements:

- Anne: There is exactly one nut among us. I did it.
- Bill: There are exactly two nuts among us. I did it.
- Caroline: There are exactly three nuts among us. I did it.

Question. *How many nuts are there among the suspects?*

Puzzle 2.5 **Nuts University**

A group of nuts works at the math department of Nuts University. All have different salaries and different workloads. Moreover, they all know each other's salaries and workloads. All of them made the following two statements, one after the other.

- There are at least 10 people in our department with a higher salary than mine.
- There are at least 10 people in our department with a higher workload than mine.

Question. *How many people work in the department?*

Puzzle 2.6 **Ten People**

In a group of 10 people, there might be knights, nuts, and knaves. They were all asked the same three questions in the same order.

1. Are you a knight?
2. Are you a nut?
3. Are you a knave?

Question. *You only know that the total number of "yes" answers out of the* $3 \cdot 10 = 30$ *responses is* X. *For which values of* X *can we guarantee that there is at least one nut in the group?*

Puzzle 2.7 **Three People**

Alice, Bob, and Carol live on the island you are visiting. One of them is a knight, one is a knave, and one is a nut. They made the following statements about their ages.

- Alice: My age is divisible by 9. I am 27. Bob is 19.
- Bob: Someone is over 20. Alice's age is even. Carol's age is odd.
- Carol: Someone is over 21. My age is even. I am 24.

Question. *Who is who, and what are their ages?*

Puzzle 2.8 **The Monastery**

Archaeologists found an ancient monastery that housed nuts and only nuts. The scripture on the walls was written in an unknown language. One theory suggests that these are mantras. A mantra is a statement that is repeatedly uttered to meditate. This theory was initially rejected, as nuts' statements must alternate between truths and falsehoods, thus repeating the same statement seems impossible. However, it was shown later on that nut mantras are possible.

Question (1). *Can you come up with an example of a nut mantra?*

Question (2). *Can you come up with a mantra that not only nuts can repeatedly utter but also knights and knaves?*

Puzzle 2.9 **What Are You?**

During your island adventure, you meet a lone vagabond.

Question. *Using a single yes-or-no question, can you determine whether the person in front of you is a knight, a nut, or a knave?*

Puzzle 2.10 **The Knight and His Numbers**

A knight thinks of a number between 1 and 100. He will answer any yes-or-no question to the best of his ability.

Question. *What is the minimum number of questions you should ask in order to uniquely determine the knight's number from his replies without guessing? (If you think you are familiar with this puzzle, think again. Read carefully.)*

Curiosity. **Pyramid Scheme**

This is the Infinite Liars Paradox. Can you determine which sentences are true and which ones are false?

All the sentences below are false.

All the sentences below are false.

All the sentences below are false.

All the sentences below are false.

All the sentences below are false.

...

Puzzle 2.11 **The Crossroads**

You feel like you've had enough of knights, nuts, and knaves, and you want to leave the island. On your way to the port, you reach a crossroads. One path will lead you to your destination, and the other will keep you lost on the island. Luckily, you encounter two islanders. The first says:

One of us always speaks the truth.

The second says:

One of us always lies.

Assume that the islanders know each other. A priori, each of them could be either a knight, a nut, or a knave, so you are not even sure whether their statements are true. Moreover, it's not even clear whether by saying "One of us..." they meant "**At least** one of us..." or "**Exactly** one of us...".

Question. *You don't want to test the patience of these strangers, so can you find your desired path by asking only one of them a single yes-or-no question?*

Solutions

This chapter was inspired by Raymond Smullyan, who invented many puzzles about knights and knaves. Such puzzles first appeared in 1978 in his book *What Is the Name of This Book?*, and continued appearing in many other books. In Smullyan's world, in addition to knights and knaves, there are normals: people who can answer truthfully or lie.

Raymond Smullyan also introduced other interesting characters. For example, in his book *The Lady or The Tiger?* we can meet sane and insane people. A sane person is correct in their beliefs, while an insane person is not. For example, an insane knave believes that 2 plus 2 is not 4; thus, being a knave, such a person would lie "I believe that 2 plus 2 is four", making such puzzles insanely hard.

Nuts appeared in many places. They are often referred to as alternators.

By the way, there is an analogy between knights and knaves puzzles and real and fake coin puzzles. There are many puzzles where you have to find fake coins using balance scales, while you are given a lot of coins that look identical. The real coins weigh the same, while the fake coins weigh the same but are lighter or heavier than the real ones. Then, the analog to nuts are special coins that change their weight each time they are put on the scale from the weight of a real coin to the weight of a fake coin and vice versa. The alternator coins were introduced by Tanya Khovanova and her students in the paper not-surprisingly titled *Alternator Coins* published in Math Horizons in 2017.

Solution to Puzzle 2.1 — Three Nuts

Suppose Anne's second statement that Bill did it alone is true, then her first statement is also true, creating a contradiction with the fact that she is a nut. Therefore, her first statement is true, and her second statement is a lie. We conclude that Anne didn't do it. One might assume that Bill did it with Caroline, but the fact that Anne's second statement is a lie means that if he did it, he didn't do it alone. It follows that either Caroline did it alone or Bill did it together with Caroline. Now, we look at Caroline's statements. We know that she definitely committed the arson. Thus, her first statement is a lie. Then, her second statement is true, and she did it alone.

Source: This puzzle was developed by the authors in 2024.

Solution to Puzzle 2.2 — Can You Repeat That?

The two statements seem completely symmetric; any reason to arrest one of the suspects could be applied to the other. The only difference is that Benjamin added "indeed" at the end of his statement, which indicates that he knew of Avigail's statements when he was uttering his own.

Let's unravel this. When considering the sentence "I don't know which of us is a nut," in isolation, we must conclude that the person stating it can't be a knave; such a person can be a knight or a lying nut. Both cases imply that the next sentence, "I didn't commit the robbery," must be true. If Benjamin's statements were independent of Avigail's statements, then we would conclude that no one committed the robbery, which is in contradiction with the fact that Radstein knew that one of them had taken part in the robbery! We must, therefore, take into consideration that Benjamin's statements were made after Avigail and, therefore, may have been created in response to her statements. But how does that change the facts?

Let us assume that Benjamin knew what day of the week it was during the interrogation. This implies that there is an extra possibility that Benjamin can figure out whether Avigail is a nut. If the day of interrogation was not a Tuesday, then Benjamin knew that Avigail lied about the day of the week. This means that whether she is a knave or a nut, her first sentence is a lie, which implies she is a nut! Benjamin, therefore, deduced that she was a nut and must have lied throughout his statements. So, we conclude that Benjamin is a knave and committed the robbery, and Avigail is a nut and didn't commit the robbery. The detective, of course, inferred all of that and arrested Benjamin.

Source: This puzzle was developed by the authors in 2024.

Solution to Puzzle 2.3 — Together or Not?

Suppose the first of Allan's statements is true. That means Allan didn't do it, and Bruno and Caio were not together, implying that only one person, Bruno or Caio, committed the burglary. Then, two of Bruno's statements are either both true or both false. Contradiction. Thus, the first of Allan's statements is a lie, and the second is true. It follows that all three of them did it together. And also, the first of Bruno's statements is a lie, and the second is true.

Source: This puzzle was developed by the authors in 2024.

Solution to Puzzle 2.4 — I Did It

Suppose there is a knight among the suspects. Then, this person committed the crime, and everyone else lied in the second statement. It follows that Caroline lied in her first statement too, implying that Caroline is a knave. Thus, there can't be more than one nut. Then,

Anne is a knight, and there is one nut. But this means that Bill lied in his first statement, and is a knave. We get a contradiction. It follows that there are no knights.

However, someone said the truth in the second statement, implying that there is at least one nut among the suspects. It follows that exactly one of the first statements is also true. It follows that there are two nuts: one said the truth in the first statement, and one lied in the first statement.

We solved the puzzle, but who is the killer? We know that Bill said the truth in his first statement, and, hence, is an innocent nut. Either Anne or Caroline is another nut and guilty.

Source: This puzzle was developed by the authors in 2024.

Solution to Puzzle 2.5 — Nuts University

Assume there were more than 20 people, say 21, for the sake of argument. Then, separate the top 10 people with the highest salaries and the top 10 people with the highest workload. This is at most 20 people, and therefore, we are left with at least one person who must be telling the truth on both statements. Since all department members are nuts, this can't happen, so we conclude there must be at most 20 people.

Now, to see that there can't be fewer than 20 people in the department, notice that if one of the top 10 people with the highest salaries also happens to be one of the top 10 people with the highest workload, then he will be lying about both statements. Therefore, when we separate the top 10 people with the highest salaries and the top 10 people with the highest workload, we must have separated at least 20 people. And non-surprisingly, 10 people with the highest workload are in the bottom 10 in their salaries.

Source: This puzzle was developed by the authors in 2024.

Solution to Puzzle 2.6 **Ten People**

Knights will say YES to the first question and NO to the other two. Knaves will say YES to the first two questions and NO to the last one. Nuts' answers depend on how they start with the first question. If they start with a lie, they will answer with three YES answers. If they start with the truth, they will have three NO answers.

If X is less than 10, then the average number of YESes per person is less than 1; thus, there has to be at least one nut. If X is more than 20, the average number of YESes per person is more than 2, and there has to be at least one nut.

Let's see what happens if we assume there are no nuts in the group. For this, define T as the number of knights and F as the number of knaves. Then, the total number of people $T + F$ is equal to 10, and the total number of YESes $T + 2F$ is equal to X. The question is, therefore, for which values of X the set of equations below admit no solution:

$$T + 2F = X,$$
$$T + F = 10,$$

where T and F are integers between 0 and 10 inclusive. Solving the above two equations for T and F, we find $T = 20 - X$ and $F = X - 10$.

Therefore, for X between 10 and 20 inclusive, the group can have $X - 10$ knaves and $20 - X$ knights and doesn't need to have a nut.

Source: This puzzle was developed by the authors in 2024.

Solution to Puzzle 2.7 **Three People**

Suppose Bob's statement, "Someone is over 20," is a lie. It follows that everyone is 20 or younger. Then "I am 27", by Alice and Carol's

"Someone is over 21" are all lies. This contradicts the fact that one of them is a knight. Thus, Bob's first statement is true, and Bob is a knight or a nut.

Whether Bob is a knight or a nut, his last statement must also be true. Thus, Carol's age is odd. It follows that the last two of Carol's statements are a lie. Thus, Carol is a knave. Hence, no one is older than 21. That means one of the ages is exactly 21. And also that the second of Alice's statements is a lie.

Hence, Alice is a nut, and Bob is a knight. It follows that Alice's age is even and divisible by 9 and not more than 21. Also, it follows that Bob is 19.

Thus, Alice is 18 and a nut, Bob is 19 and a knight, and Carol is 21 and a knave.

Source: This puzzle was developed by the authors in 2024.

Solution to Puzzle 2.8 **The Monastery**

Solution to question 1:
One possible nut mantra is

My last statement was a lie.

You may be able to come up with more examples. Some people say that using this mantra, nuts try to convince others that they are actually sane.

Solution to question 2:
Two possible mantras that knights, nuts, and knaves can repeatedly utter are

Two statements ago, I spoke the truth

or

This statement is true.

While the second mantra can be consistently assigned either true or false values, one must always be careful when using self-referencing logics to determine the truth value of a statement. Take a look at the curiosity that follows ("Curry's Paradox") to understand why.

Source: This puzzle was developed by the authors in 2024.

Curiosity. **Curry's Paradox**

If this statement is true, this book will become a bestseller.

When considering the statement above, there are four options:

1. The statement is true, and the book will become a bestseller.
2. The statement is true, and the book will not become a bestseller.
3. The statement is false, and the book will become a bestseller.
4. The statement is false, and the book will not become a bestseller.

The statement claims that option 2 can't occur. Therefore, if it is false, option 2 occurs. But option 2 claims that the statement is true, which leads to a contradiction. We are therefore forced to conclude that the statement must be true. This furthermore implies that: *This book will become a bestseller.*

Solution to Puzzle 2.9 **What Are You?**

One possibility is asking the following question:

What would you have answered if you were asked at a random time of the day: "Would you say that 1 + 1 = 2?"?

Knights would reply with "Yes". Knaves would reply with "No". Nuts would not reply to this question since, given the condition, "at a random time of the day" makes it impossible for the nut to know the truth value of their statement. A breakdown of why that is can be seen in the following table.

Question	Knight	Knave	Nut (truthful)	Nut (liar)
Is $1 + 1 = 2$?	Yes	No	Yes	No
Would you say that $1 + 1 = 2$?	Yes	Yes	Yes	Yes
What would you answer if you were asked: "Would you say that $1 + 1 = 2$?"?	Yes	No	Yes	No
What would you have answered, if you were asked at a random time of the day: "Would you say that $1 + 1 = 2$?"?	Yes	No	?	?

The "random time of the day" part of the question doesn't affect knights or knaves.

Source: This puzzle was developed by the authors in 2024.

Solution to Puzzle 2.10 The Knight and His Numbers

If you found the number 7, you are on the right track. Try to find out why the right answer is instead 5.

First, let's see why seven seems right: After every question, the knight answers honestly yes or no. Therefore, we can divide the set of possible answers into two sets as evenly as possible. For example, we can ask whether his number is less than or equal to 50. After one such question, we are left with at most 50 possibilities, and then 25, and then 13, . . ., and finally down to 1 after seven questions.

Now, how could the answer possibly be five? First, let us realize that if we somehow could divide the set of 100 numbers into three groups per question rather than two, then we could narrow down the options to at most 34 after one question, followed by 12, and then 4, and 2 and finally 1 in five questions. To divide the set into three groups per question, we simply need to craft clever questions such that "yes" and "no" are only appropriate for two-thirds of the set of possibilities, and the remaining one-third is impossible to answer. Here is one example of such a question for the first round (though many other question-crafting strategies exist):

"Function $f(x)$ is equal to 1 for all values of x between 1 and 34 inclusive, it is equal to -1 for all values of x between 35 and 67 inclusive, and no one knows what $f(x)$ is for other values of x. Tell me, when x is your number, is $f(x)$ positive?"

P.S.: If you want to craft other types of tricky questions like this, one way is to use well-known mathematical functions that are not well defined over a certain region (e.g., $1/x$ is undefined for $x = 0$).

Source: This solution was developed by Ivo David, Caio Viana, Maury Oliveira, and Marcus Moreira in 2012 in response to the classical 7 answer solution. Independently, Tanya Khovanova published a miniature version of this technique in her blog in 2024 https://blog.tanyakhovanova.com/2024/02/guess-the-number-in-one-question/.

Solution to Puzzle 2.11 The Crossroads

This one is inspired by the classic logic puzzle in which you *know* that one of the islanders always lies, and the other always speaks the truth. In the classic version, to find your path, you can ask one of the islanders: Where would your friend tell me I should go? Whether you ask the knight or the knave of that version, they will both point in the wrong direction, so you should choose the other path.

In our version, however, you are not sure who stands in front of you. To start, let us begin by assuming that by "One of us..." the islanders meant "**At least** one of us...". The second islander, who said that at least one of them always lies, can't be a knave, as that would mean his statement is true, which contradicts the knave ethos. He also can't be a knight, as that would mean the first speaker is a knave who speaks the truth. So the second islander must be a nut. We find that there are only three options for these islanders:

Option #	**Islander 1**	**Islander 2**
1	knave	truthful nut
2	knight	lying nut
3	lying nut	lying nut

If by saying "One of us..." the islanders meant "**Exactly** one of us...", an additional option opens up:

Option #	**Islander 1**	**Islander 2**
4	knave	knave

Unlike the classical version, where you know that one of the islanders will lie to you and the other will speak the truth, here you know that, replying to the next question, either whoever you ask will lie to you, or whoever you ask will speak the truth. In options 1 and 4, you will hear a lie for the next question, and in options 2 and 3, you will hear the truth. Since you don't know in front of which of these options you stand, you can use the same question as in the classical riddle and ask one of the islanders:

Where would the other islander tell me I should go?

However, in this version of the puzzle, you should follow their instruction and walk that path!

Source: This puzzle was developed by the authors in 2025.

Chapter 3

Unusual Numbers

If mathematics is the language of the universe, then unusual numbers are the tongue twisters. Unlike your everyday number property, here, we hope you will find some of the more unconventional variety... For example, did you know that 1 and 8 are the only cubes in the Fibonacci sequence? Or that 5 is the smallest number of queens needed to attack every square on a standard chessboard? Or even better, that 4 is the only number for which the number of letters in its English name **four** is equal to the number itself!?

The challenges in this chapter reveal hidden surprises within numbers, equations, and mathematics in general, inviting you to think outside the box. Whether you're discovering curious linguistic connections or exploring uncommon numerical phenomena, you'll find here a fresh perspective on numbers. Get ready to uncover the extraordinary in the ordinary.

Puzzle 3.1 **Five Has One Syllable**

What is the only number between 1 and 100 with five syllables in its name?

Puzzle 3.2 **Prime Reversal**

What is the smallest prime number whose reversal is a composite (not a prime) number?

Puzzle 3.3 **Divisible by 24**

Can you prove that if p is a prime number greater than 3, then $p^2 - 1$ is divisible by 24?

Curiosity. **The Number 315 Is Not Important**

Someone once said "If it doesn't have a Wikipedia page, it's not important". As of the time of writing of this book, the number 315 is the smallest natural number that is not important. We believe this property makes it quite important.

Puzzle 3.4 **Unalphabetical Numbers**

What is the only number written with all letters in alphabetical order? What is the only number written with all letters in reverse alphabetical order?

Puzzle 3.5 **Find B**

What is the smallest number that uses the letter B?

Puzzle 3.6 **Number Dictionary**

An infinite dictionary contains all integer numbers in alphabetical order, written in English. What is the first number in this dictionary? What is the last number?

Puzzle 3.7 **Unusual Equality**

Find x and y greater than 0 such that $x^x = y^y$ but $x \neq y$.

Curiosity. **A Forbidden Equation**

The three most beloved mathematical constants are the golden ratio $\phi = 1.61803\ldots$, Euler's number $e = 2.71828\ldots$, and the mathematical constant $\pi = 3.14159\ldots$ However, be careful never to show your math teacher the following forbidden equation:

$$\frac{\phi \cdot e}{\pi} = \frac{7}{5}.$$

(But if you do, for your own safety, don't forget to add 0.00001358... to the right side of the equation [5].)

Puzzle 3.8 **Primes Followed by a Square**

Can you find a prime number followed by a square number? How many such primes are there?

Puzzle 3.9 **Primes Followed by a Cube**

Can you find a prime number followed by a cube number? How many such primes are there?

Curiosity. **Fibonacci Divisibility**

Have you ever heard of the following property? For any integer n, there exists a Fibonacci number divisible by n. It makes the following property amusing: No odd Fibonacci number is divisible by 9.

Puzzle 3.10 **More than Numbers**

Consider this equation: $x^4 - x^2 + y^2 = 0$.

Question. *How big is it?*

Puzzle 3.11 **Palindromic Prime**

A number is palindromic if it is invariant to reversing its digits; e.g., if you reverse the digits of the number 13431, you get 13431; thus, by definition, 13431 is palindromic.

Question. *Find a palindromic prime with an even number of digits. Prove that the answer is unique.*

Puzzle 3.12 **Broken Calculator**

We have a strange calculator that seems to be broken, but it is not. It follows its own rules. Our calculator can add up any sequence of one-digit numbers:

It says that:

I. $7+5=9$	IV. $9+7=9$
II. $1+4=4$	V. $4+6=8$
III. $0+2=8$	VI. $2+2=2$

Question. *Playing with my calculator, I found* $X \neq 4$ *such that* $X+3=9$ *and* $X+7 \neq 9$. *Find* X.

Puzzle 3.13 **Math Is Broken**

Johnny's homework was to calculate $5548-5489$. He typed it into his calculator and found that the result was 59. He tried to remember what they did in the classroom, and he reasoned that 548 had to cancel out, and he was left with 59:

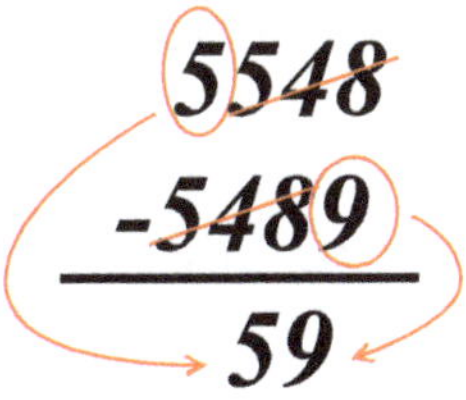

To test his technique, he tried once again typing in two numbers $XXYZ - XYZW$, different from before, where X, Y, Z, and W are distinct digits, and once again, he found that it was indeed equal to XW!

Question. *Considering the new values of X, Y, Z, and W, which digits are the same (identical and in the same location) as before?*

Puzzle 3.14 **Cheapest Trip to Space**

Here is the largest denomination ever printed on a banknote. It was issued by the Bank of Hungary in 1946.

It reads "SZÁZMILLIÓ B.-PENGŐ", which translates to One Hundred Million Trillion Pengő (the B stands for Billió in Hungarian, which equates to a million million, which equals a trillion, and pengő is the name of the currency).

Question. *Given that the (i) altitude of the International Space Station is 418 km, the (ii) distance to the Moon is $3.6 \cdot 10^5$ km, the (iii) distance to the Sun is $1.4 \cdot 10^8$ km, and the (iv) distance to the nearest star to our solar system, Proxima Centauri, is $4 \cdot 10^{13}$ km,* *if you were to exchange this bill for a stack of coins, each worth one pengő, how far would this stack of coins go?*

Curiosity. **No Money!**
Did you know that you can actually buy 0 euro banknotes?

Question. *How much would you be willing to pay for a 0 euro banknote?*

One of the authors once bought a 0 euro bill for 10 euros.

Solutions

As we close this chapter, let's take a moment to reflect on this delightful collection of curiosities and mathematical oddities. Each puzzle, unlike many boring classroom activities, tells a unique story and reveals that even ordinary-looking numbers can hide delightful secrets and contain unique enchanting colors.

One story that highlights this tale is that of the Hardy–Ramanujan number:

$$1729.$$

The story goes that Hardy, while visiting his student, the Indian mathematician Ramanujan, in the hospital, mentioned disparagingly that the license plate of the taxi that brought him there had, unfortunately, a "dull" number: 1729. He added, "Hopefully, it is not an unfavorable omen", to which Ramanujan replied: "It is a very interesting number!" since "it is the smallest number expressible as the sum of two cubes in two different ways![1]"

Just as Ramanujan's creativity and mathematical artistry were capable of revealing the magic of such a "dull" number, we hope that you will be encouraged by this chapter to have your vision of numbers, even the dull ones, reinchanted by wonder. Mathematics is not just about equations and calculations; it's about discovery and finding out what is special about every number and letting every puzzle tell you a story. You might just be staring at a beautiful gem of mathematics, like 1729, and all it takes is just one more step to find out!

For more unusual properties, mathematical tales, and numbers galore, we invite you to visit Tanya Khovanova's Number Gossip website at `www.numbergossip.com`. There, you can find many more secrets, hidden connections, and properties that are rarely discussed elsewhere. And, if you are feeling bold, how about you submit a new and unknown property to this wonderful encyclopedia of oddities?

---

[1]The two ways are: $1729 = 10^3 + 9^3 = 12^3 + 1^3$.

Solution to Puzzle 3.1 **Five Has One Syllable**

The answer is 77: se-ven-ty-se-ven.

Source: This is a puzzle adaptation of a well-known number property.

Curiosity. **Persistence**
Did you know that 77 is the smallest number with multiplicative persistence 4? Multiplicative persistence denotes the number of times you multiply the digits of the number, repeating with the product until a single digit is obtained. So, for 77, we multiply 7 times 7 to get 49, then 4 times 9 to get 36, then 3 times 6 to get 18, then 1 times 8 to get 8, and stop. We used multiplication four times. Thus, the multiplicative persistence of 77 is 4.

Solution to Puzzle 3.2 **Prime Reversal**

One-digit prime numbers reverse to themselves. Therefore, our number has more than one digit. Checking a few prime numbers, we get our answer: 19. Reversing its digits gives 91, which might look like a prime at first glance, but $91 = 7 \cdot 13$. An image of 19 made by Ottilia Adelborg is depicted below.

Source: Folklore.

Solution to Puzzle 3.3 — Divisible by 24

The expression $p^2 - 1$ decomposes as $(p-1)(p+1)$. These are two consecutive even numbers. Therefore, one of them is divisible by 4, and their product is divisible by 8. In addition, since one out of three consecutive numbers must be divisible by 3, we know that one of $p-1$, p, or $p+1$ is divisible by 3. But since p is a prime greater than 3, it is not divisible by 3, and therefore either $p-1$ or $p+1$ is divisible by 3. We see that $p^2 - 1$ is divisible by 8 and by 3, which means it is divisible by 24.

Source: Folklore.

Solution to Puzzle 3.4 — Unalphabetical Numbers

The only number written with all letters in alphabetical order is forty, and the only number written with all letters in reverse alphabetical order is one.

Source: This is a puzzle adaptation of well-known number properties.

Solution to Puzzle 3.5 — Find B

The smallest number that uses the letter B is one billion.

Source: This is a puzzle adaptation of a well-known number property.

Solution to Puzzle 3.6 — Number Dictionary

The first number in the dictionary is eight, and the last is zero.

Source: This is a puzzle adaptation of well-known number properties.

Bonus. How can an infinite dictionary contain both the first and last items?

Solution to Puzzle 3.7 **Unusual Equality**

One simple answer is 1/2 and 1/4; however, there are many others. It might feel strange that such an equality is possible, but unlike what one might think, x^x is not an increasing function! It is decreasing from 0 and has a minimum at $1/e$, after which it is increasing.

Source: Folklore.

Bonus. The function $f(x) = x^x$ is not an increasing function, but is $f(x) = x^{x^x}$?

Solution to Puzzle 3.8 **Primes Followed by a Square**

Suppose the square in question is n^2, then our prime is n^2-1, which decomposes to $(n-1)(n+1)$. As our number is prime, we have $n-1=1$, implying that $n=2$, and our prime number is 3. There is only one such prime number.

Source: Folklore.

Solution to Puzzle 3.9 **Primes Followed by a Cube**

Suppose the cube in question is n^3, then our prime is n^3-1, which decomposes to $(n-1)(n^2+n+1)$. As our number is prime, we have $n-1=1$, implying that $n=2$, and our prime number is 7. There is only one such prime number. A beautiful image of number 7 is made by Ottilia Adelborg and is depicted below.

Source: Folklore.

Solution to Puzzle 3.10 **More than Numbers**

The figure below shows the plot of the equation, which resembles the infinity sign. Thus, infinity is the answer.

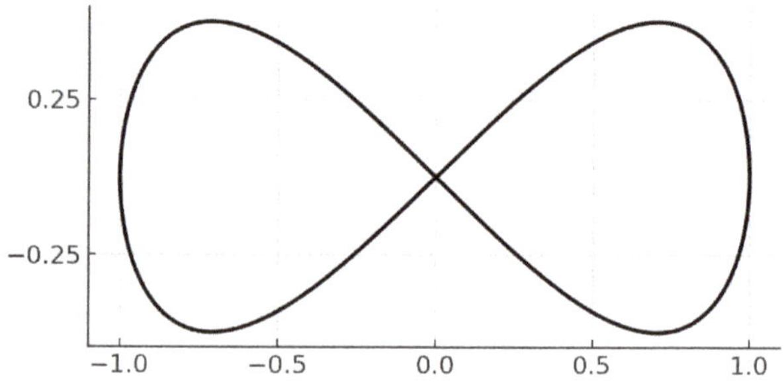

Source: Folklore.

Solution to Puzzle 3.11 — Palindromic Prime

A palindromic prime with an even number of digits is divisible by 11. To see this, remember the 11 divisibility rule: A number is divisible by 11 if and only if the alternating sum of digits is divisible by 11. The alternating sum of digits is the sum of digits in odd-indexed places minus the number of digits in even-indexed places. For example, the alternating sum of digits of 1234 is $1-2+3-4=-2$. We leave it to the reader to check that the alternating sum of digits of a palindrome with an even number of digits is 0. Thus, such a number is divisible by 11. To be prime, it has to be 11.

Source: Folklore.

Solution to Puzzle 3.12 — Broken Calculator

The digits in this calculator are represented in a seven-segment display. When the calculator adds two numbers, it doesn't do any arithmetic calculations. Instead, it looks at their segment representation and checks for each segment separately whether it is ON in at least one of the summed numbers. If so, it will be ON in the result, and it will be off otherwise. Therefore, when adding two single-digit numbers $A+B$ there is no guarantee that the result will even be a number. It might actually turn out to be some strange collection of segments turned on.

Let us then look at all possible sums we can form with the digits 0 to 9.

$0+0=0$	$0+1=0$	$0+2=8$	$0+3=8$	$0+4=8$	$0+5=8$	$0+6=8$	$0+7=0$	$0+8=8$	$0+9=8$
	$1+1=1$	$1+2=Y$	$1+3=3$	$1+4=4$	$1+5=9$	$1+6=8$	$1+7=7$	$1+8=8$	$1+9=9$
		$2+2=2$	$2+3=Y$	$2+4=8$	$2+5=8$	$2+6=8$	$2+7=Y$	$2+8=8$	$2+9=8$
			$3+3=3$	$3+4=9$	$3+5=9$	$3+6=8$	$3+7=3$	$3+8=8$	$3+9=9$
				$4+4=4$	$4+5=9$	$4+6=8$	$4+7=Z$	$4+8=8$	$4+9=9$
					$5+5=5$	$5+6=6$	$5+7=9$	$5+8=8$	$5+9=9$
						$6+6=6$	$6+7=8$	$6+8=8$	$6+9=8$
							$7+7=7$	$7+8=8$	$7+9=9$
								$8+8=8$	$8+9=8$
									$9+9=9$

Note that there are two unusual symbols produced with the sums of the digits 0 to 9: $1+2=2+3=2+7=Y$ and also $4+7=Z$. This is what these strange additions look like: $Y =$  and $Z =$ 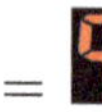. All that is left is to verify what sums these strange symbols produce:

$Y+0=8$	$Y+1=X$	$Y+2=Y$	$Y+3=8$	$Y+4=8$	$Y+5=8$	$Y+6=8$	$Y+7=Y$	$Y+8=8$	$Y+9=8$
$Z+0=8$	$Z+1=Z$	$Z+2=8$	$Z+3=9$	$Z+4=Z$	$Z+5=9$	$Z+6=8$	$Z+7=Z$	$Z+8=8$	$Z+9=9$

Oh yes, and don't forget to check that $Y + Z = 8$, and this way, we can definitely conclude that no more new symbols are created by addition! Thus, we conclude that X must be equal to $4 + 7 = Z$.

Source: This puzzle was developed by the authors in 2024.

Solution to Puzzle 3.13 — Math Is Broken

The answer is: Z and W must be 8 and 9 just as before.

We have $1100X+10Y+Z-1000X-100Y-10Z-W = 10X+W$. It follows that $90X-90Y = 9Z+2W$. Therefore, W must be divisible by 9, which means it is either 0 or 9. Also, $9Z+2W$ must be divisible by 10.

Suppose $W = 0$, then $Z = 0$, which contradicts the digits being distinct.

Suppose $W = 9$, then $9Z + 18$ is divisible by 10, which means $Z = 8$. We are left with the following equation $90X - 90Y = 90$, implying that $X = Y + 1$. Thus, Z and W have to be the same, and X and Y can be something else.

Source: This puzzle was developed by the authors in 2024.

Solution to Puzzle 3.14 — Cheapest Trip to Space

Let's unravel the number. A trillion is one thousand billion, which is 10^{12}, one hundred million is 10^8, thus, the bill is worth 10^{20} Pengő! If the distance to Proxima Centauri were divided into 10^{20} equal steps, each step would need to be 0.4 mm in length only. Even the smallest and thinnest coin would be sufficient to reach the nearest star and come back!

Source: This puzzle was developed by the authors in 2024.

Chapter 4

Dicey Dice

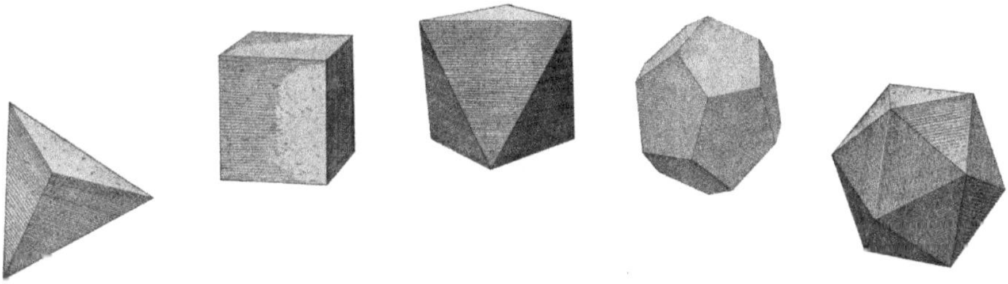

People have been using dice in games for more than 4000 years. However, isn't it interesting that some of their properties have remained hidden all this time? In this chapter, you'll find puzzles that showcase some of these features that have only recently been discovered, including quirky, overlooked, and paradoxical aspects of these wonderful, playful toys we call dice. For example, many games, such as Settlers of Catan, care only about the sum of two standard dice rolls, and the most frequent value for the sum in this case is 7, happening once in every six rolls. George Sicherman discovered a non-standard pair of six-sided dice whose sums have the same probability distribution as a pair of standard dice. This pair of dice is numbered 1, 3, 4, 5, 6, 8 and 1, 2, 2, 3, 3, 4, and is unique! Can you check that the probability of the sum being 7 is indeed $\frac{1}{6}$?

To be a bit more precise, it is important to know that throughout this chapter, we will call any random number generator that produces the numbers 1 to 6 with a uniform probability a $D6$; this includes but is not limited to the standard six-sided die. Similarly, we call a random number generator that produces the numbers 1 to N with uniform probability a DN. Also, here, dice will almost always be associated with specific geometric shapes: $D4$s are usually in the shape of a tetrahedron (a pyramid-like shape with a triangle base), a $D6$ is a cube, a $D8$ is an octahedron, and so on and so forth... but it doesn't need to be that way. In fact, "non-Platonic shapes" like mysterious trapezohedrons and other wild geometric shapes can also have symmetries that ensure that landing on every side is equally probable. Even though this chapter is not about geometry, it might be useful to know the five basic Platonic solids illustrated above: a tetrahedron, a cube, an octahedron, a dodecahedron, and an icosahedron.

In this chapter, we visit a club of dice collectors, where the atmosphere is filled with excitement. Enthusiasts eagerly share their brilliant ideas for new dice designs, each with their own unique favorites. Some of the dice collectors might instead show coins rather than 3D geometric objects! These ones are a little bit odd among the dice enthusiasts, but are still a welcome part of the family. In our notation, coins are also considered dice and are, therefore, typically denoted as $D2$s. As we move from one collector to another, they share their passion for these small yet remarkable objects, making our experience truly unforgettable.

Puzzle 4.1 **Four-Sided *D*6**

Takahashi shows you a tetrahedron-shaped die with faces labeled with 0, 1, 2, and 4 pips, respectively. He outlandishly claims it is a $D6$!

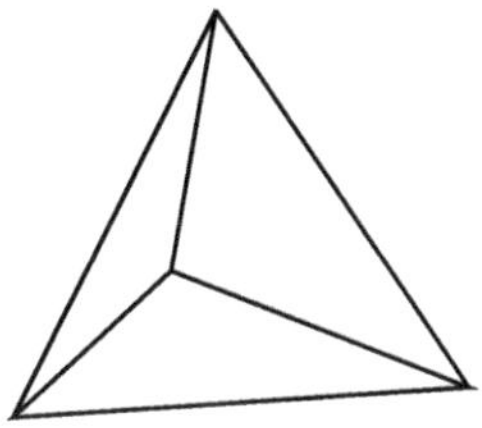

Question. *How can it be that this four-sided die is a* $D6$*?*

Puzzle 4.2 **Unfair Advantage**

At the club, Ernesto challenges anyone to a one-time coin-flipping challenge: he flips the coin n times, and the opponent flips a coin $n + 1$ times; the one with the most heads wins. Adil takes up the challenge.

Question. *What is the probability that Adil will win?*

Puzzle 4.3 **Vertex Dice**

Felix has a cube-shaped die on display: It has six sides, eight vertices, and twelve edges, just like any other cube. When he rolls his die, he identifies the closest vertex to him and adds the number on the three faces that share that vertex. He claims it is a $D8$!

Question. *Can you assign number values to each face so that you produce a $D8$ this way?*

Curiosity. **Calendar Dice**

Have you heard about calendar dice? It is a pair of six-sided dice that can enumerate all the days of the month: 01, 02, ..., 30, 31. Looking at the first day of the month, it is clear that one die has to carry a face value of 0, and the other die must carry a face value of 1. Can you find the other faces of such a pair?

Hint: *It is impossible to do so with pips, but it is possible with numerals.*

Puzzle 4.4 — Edge Dice

Ella has a cube-shaped die: It has six sides, eight vertices, and twelve edges. When she rolls her die, she identifies the closest edge to her and adds the numbers on the two faces that share that edge.

Question. *Can you assign number values to each face so that you produce a $D12$ this way?*

Puzzle 4.5 — Subtraction Dice

Geliy has a cube: It has six sides, eight vertices, and twelve edges. When he rolls a cube-shaped die, he reads the top face followed by the face on the side pointing to him. Now, instead of addition, he will subtract the top number from the side number.

Question. *Can you assign number values to each face so that he would produce the numbers -12 to -1 and the numbers 1 to 12?*

Puzzle 4.6 **More Edge Dice**

Inna has an octahedron-shaped die: It has eight sides, six vertices, and twelve edges. When she rolls the die, she identifies the closest edge to her and adds the numbers on the two faces that share that edge.

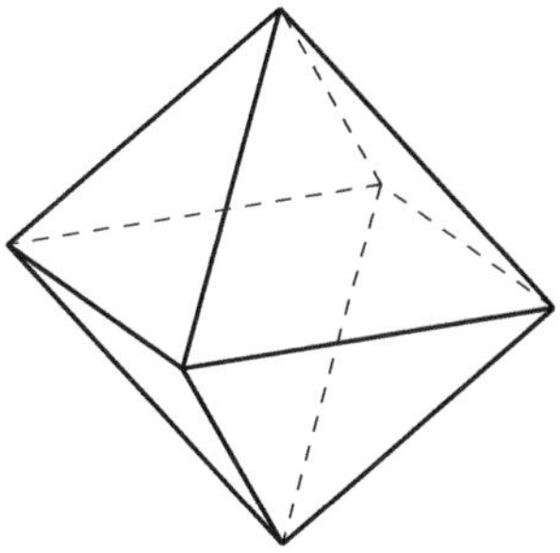

Question. *Can you assign number values to each face so that she would produce a D12 this way?*

Curiosity. **Independent Events**

Here is a seemingly innocent scenario: you roll two $D6$s and add up the results. Now, consider the following probabilistic events:

Event A. Your first die roll is 3.
Event B. Your added total is 7.
Event C. Your added total is 8.

Now, see if you can come to terms with this fact: Events A and B are independent[a]; however, A and C are not.

[a]In probability theory, two events A and B are considered independent if the probability that A happens is not affected by whether or not B has happened and vice-versa.

Puzzle 4.7 **Coin Bet**

Alice and Bob are playing the following game. Alice will toss a fair coin until someone wins. Alice bets that the sequence of two heads in a row will appear first. Bob bets that the sequence tails-heads will appear first.

Question. *Who has the advantage?*

Puzzle 4.8 **Maxing Out Your Dice**

Suppose you have five standard Platonic solid dice: $D4$, $D6$, $D8$, $D12$, and $D20$. Suppose you roll all of them.

Question (1). *What is the probability that the sum is 50? What is the probability that the sum is 49?*

Question (2). *What is the probability that the product of the numbers you roll is 46080? (Note that* $46080 = 4 \cdot 6 \cdot 8 \cdot 12 \cdot 20$*.) What is the probability that the product of the numbers you roll is 23040?*

Puzzle 4.9 **Another Dice Probability**

Suppose Alice has two $D6$s and Bob has a $D4$ and a $D8$. They both roll their dice.

Question. *What is the probability that Bob's sum is greater than Alice's sum?*

Curiosity. **Lake Wobegon Dice**

Have you heard of Lake Wobegon Dice? Discovered by Jorge Moraleda and David G. Stork in 2012, this strange set of dice has the property that every die "rolls higher than the set average":

$$\text{Die } 1 = \{6, 6, 6, 6, 6, 6\}$$
$$\text{Die } 2 = \{1, 1, 7, 7, 7, 7\}$$
$$\text{Die } 3 = \{4, 4, 4, 7, 7, 7\}$$

More precisely, in a single three-dice roll, the probability that Die n is larger than (Die 1 + Die 2 + Die 3)/3 is greater than 1/2 for every $n = 1, 2, 3$.

Source: Moraleda, J., & Stork, D. G. (2012). Lake Wobegon Dice. *The College Mathematics Journal*, 43(2), 152–159.

Lesson. Next time you hear mothers bragging about how great their kids are in school, and at some point in the conversation, each mother says: "My child is above average!". Who would have guessed that maybe they are all right!

Solutions

Dice are fun, but aren't they also mysterious? They somehow embody both the excitement of surprise as well as the chaos of unpredictability. They are both toys that date back thousands of years, but also, they are tools used in divination by mystics and sages, the theme of one of our next chapters.

In this chapter, however, we have chosen to talk exclusively about dice both because we love dice and also because, in many ways, dice are a mathematician's best souvenir. For the geometer, they are Platonic solids, perfect objects with coveted symmetries in the world of three-dimensional shapes; for the probabilist, they are random number generators, exactly the type of engine one needs to run a probability experiment; and for the algebraist, they contain the seeds to the natural numbers, that is, the root of number theory!

So many more interesting dice properties have been written about in the literature that we just could not fit into this book! For the curious-minded, if you are thirsty for more dice topics not covered in this book, you might want to look into: (i) *Intransitive Dice* [6], (ii) *David vs Goliath Dice* [7], and (iii) *Go-First Dice* [8]; which are many a dice collector's favorite.

We hope that this short dive into the world of dice will warm your heart to these bite-sized, math-crunchy, gem-like, random number generators!

Solution to Puzzle 4.1 — Four-Sided *D6*

A tetrahedron has six edges. When Takahashi rolls the die, he identifies the closest edge to him (rather than the closest face) and adds the number on the two faces that share that vertex. Here is what the rolls 1 to 6 look like:

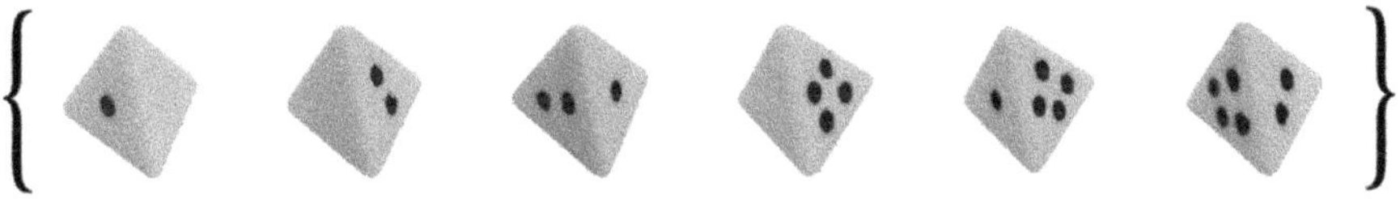

Source: This die was invented by David in 2023.

Solution to Puzzle 4.2 — Unfair Advantage

Consider $n+1$ tosses. They either produce more heads or more tails than n tosses. It can't be that they produce both more heads and more tails, as otherwise, the total number of tosses is more than $n+1$. Therefore, by symmetry between heads and tails, the probability that Adil with $n+1$ tosses will beat Ernesto with n tosses is $1/2$. Below, you can see a breakdown of all the possible options for $n = 1$. Out of 8 possible outcomes, Adil wins 4, draws 3, and loses 1.

	TT	**TH**	**HT**	**HH**
T	Tie	Adil Wins	Adil Wins	Adil Wins
H	Ernesto Wins	Tie	Tie	Adil Wins

Source: Folklore.

Solution to Puzzle 4.3 — Vertex Dice

The answer is yes. And there are multiple solutions; here is one: Assign the numbers 0, 0, 1, 2, 2, and 4 to the faces of your die in the following configuration:

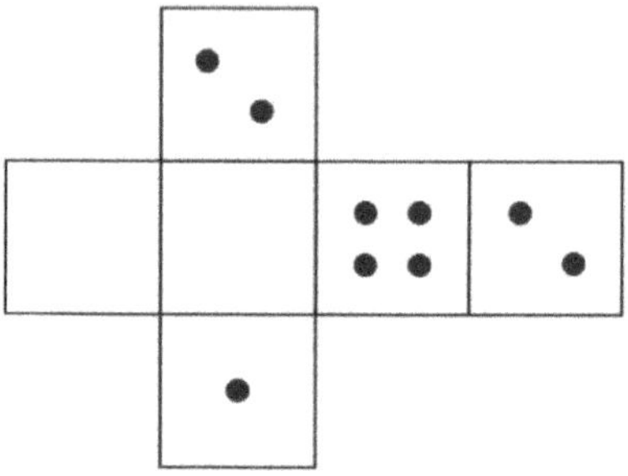

The rolls 1 to 8 look like this:

Source: This die was invented by David in 2023.

Solution to Puzzle 4.4 — Edge Dice

The answer is, unfortunately, no!

Suppose there is a solution. Then, the sum of all the visible sums is $1+2+3+\cdots+12=78$. On the other hand, each face participates in four sums. Therefore, the face values must add up to $78/4=19.5$. This immediately shows us that the face values can't all be integers. Thus, there exists a face with value $X+Y$, where X is an integer and $0<Y<1$. Therefore, all its neighbor faces have the non-integer component equal to $1-Y$. However, you can find two neighbors of this face that are neighbors of each other. Thus, $2(1-Y)$ must be an integer, and $Y=0.5$. Repeating the previous argument for other faces, we get that all faces must have integer $+0.5$ values. Six integer values plus six halves add up to an integer. Therefore, they can't add up to 19.5.

Source: This result is by David in 2024 with some helpful discussion from Rod Bogart.

Curiosity. **Yes Yes Yes ... Yes No!**
Even though the puzzle above does not admit a solution, there are some quite interesting and annoying near-solutions. Here we show our favorite one that produces the numbers 1 to 11 and then 14.

How annoying is that?!

Solution to Puzzle 4.5 — Subtraction Dice

The answer is yes! And there are multiple solutions. Here is one configuration:

	12		
10	0	8	17
	11		

This is how the rolls 1 to 12 are obtained:

$$\begin{array}{cccc} 1 = 11 - 10 & 2 = 12 - 10 & 3 = 11 - 8 & 4 = 12 - 8 \\ 5 = 17 - 12 & 6 = 17 - 11 & 7 = 17 - 10 & 8 = 8 - 0 \\ 9 = 17 - 8 & 10 = 10 - 0 & 11 = 11 - 0 & 12 = 12 - 0. \end{array}$$

Source: This die was invented by David in 2024.

Bonus. Can you assign face values to a tetrahedron-shaped die such that it produces the numbers -6 to -1 and 1 to 6 using the technique above?

Solution to Puzzle 4.6 **More Edge Dice**

The answer is yes! Here is one solution:

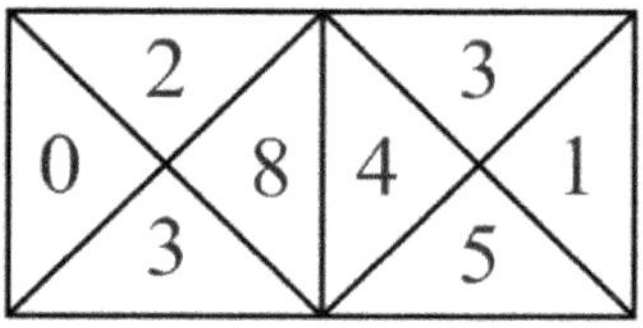

The rolls 1 to 12 look like this:

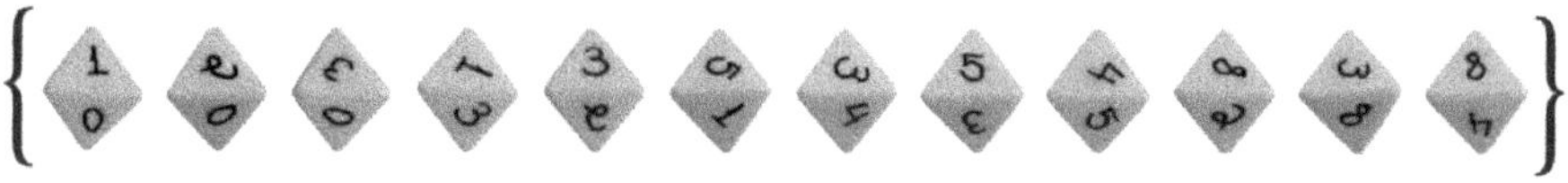

Source: This die was invented by David in 2024.

Solution to Puzzle 4.7 **Coin Bet**

If the first toss is tails, then Bob wins. Indeed, the next heads will give Bob the winning sequence without Alice having a chance. If the first toss is heads, then Alice wins only if the second toss is heads, too. If the second toss is tails, then again, the win is in Bob's hands. Thus, Alice wins with probability $\frac{1}{4}$, while Bob wins with probability $\frac{3}{4}$.

Source: Folklore question related to Penney's game.

Curiosity. **Penney's Game**

In Penney's game, Alice selects a sequence of three heads and tails and shows this sequence to Bob, for example, HHH. Bob then selects another sequence of heads and tails of the same length, for example, THH. After that, a fair coin is tossed. The player whose sequence appears first wins. Surprisingly, despite Alice having the first choice, Bob always has the advantage.

Source: This game was first studied by Walter Penney and published in the *Journal of Recreational Mathematics* in 1969.

Solution to Puzzle 4.8 — Maxing Out Your Dice

The answer to the first part of both the first and second questions is 1/46080. To the second part of the first question: one of the dice is one less, so we choose which one in five ways, implying that the answer is 5/46080. The product is different from the sum, and the answer is counterintuitive. Similar to before, we can choose a die, say D6, and divide the maximum value by 2 and get the set of values 4, 3, 8, 12, and 20 for dice $D4$, $D6$, $D8$, $D12$, and $D20$, correspondingly. However, we can get the same values if we swap the values in the $D4$ and the $D6$: 3, 4, 8, 12, 20. More unexpectedly, we can produce the same result with 3, 5, 8, 12, and 16, and 5 is not half of the maximum on any die.

To calculate the total, we need to find five numbers a, b, c, d, and e such that $a \leq 4$, $b \leq 6$, $c \leq 8$, $d \leq 12$, $e \leq 20$, and $abcde = 23040$. There are 22 different ways to roll the dice and get this total. Thus, the answer is 22/46080.

If you are wondering about all the possibilities, here is the list:

$$(2,6,8,12,20),\ (3,4,8,12,20),\ (3,5,8,12,16),\ (3,6,8,8,20),$$
$$(3,6,8,10,16),\ (4,3,8,12,20),\ (4,4,6,12,20),\ (4,4,8,9,20),$$
$$(4,4,8,10,18),\ (4,4,8,12,15),\ (4,5,6,12,16),\ (4,5,8,8,18),$$
$$(4,5,8,9,16),\ (4,5,8,12,12),\ (4,6,4,12,20),\ (4,6,5,12,16),$$
$$(4,6,6,8,20),\ (4,6,6,10,16),\ (4,6,8,6,20),\ (4,6,8,8,15),$$
$$(4,6,8,10,12),\ (4,6,8,12,10).$$

Source: This puzzle was developed by the authors in 2024.

Solution to Puzzle 4.9 — Another Dice Probability

Suppose Alice rolls (a, b), and Bob rolls (c, d). Alice wins when $a+b > c+d$ and loses when $a+b < c+d$. The probability of this roll is the same as the probability of the roll $(7-a, 7-b)$ by Alice and $(5-c, 9-d)$ by Bob. The sum for the roll $(7-a, 7-b)$ is $14-(a+b)$, and the sum for the roll $(5-c, 9-d)$ is $14-(c+d)$. The new roll reverses the winner. Therefore, they have an equal chance of winning.

Source: This puzzle was developed by the authors in 2024.

Chapter 5

Ants and Their Ways

Ants. Tiny little creatures that somehow make their way through this complex world of ours. A world, no doubt, surrounded by giants of disproportionate dimensions, in which their tiny little antennae are barely adequate to grasp. Can you imagine what a 1.7 m tall human looks like in proportion to a 2 mm ant? In height, the tallest building in the world[1], which has 163 floors and is 828 m tall, would need an additional 121 floors (or 617 m) to produce a height proportionate for that same person! In weight, however, even two copies of this building would not be enough to simulate how heavy a human is compared to the poor ant. We would need 14 of these buildings!

However, don't feel sorry quite yet for these tiny six-legged friends of ours. One group of crazy scientists recently estimated that there are currently 20 quadrillion ants on Earth [12]. This means they outnumber us by more than 2.5 million to one! So ... maybe, just maybe, they know something we don't.

In this chapter, we will delve into this world of tiny ants. Ants, for sure, can't waste time with those tiny legs, and so, in this chapter, you can expect to see "random walking" ants hopelessly exploring the world but also "hive-minded" ants blindly following the colony order into peril. Both of these raise interesting questions for the

[1] The Burj Khalifa built in Dubai in 2010 currently holds the world record for tallest man-made structure!

curious-minded. We hope to shed some light on this cute corner of the world and make you think like an ant with us for a short and refreshing chapter.

Just one clarification before we begin: Mathematicians like using ants in their puzzles because ants are so tiny that their size is negligible. Mathematicians assume that an ant is a dot, and so do we throughout this chapter.

. . .

Puzzle 5.1 **Ants on a Wire**

We open this chapter with a classic puzzle. One hundred tiny ants are randomly placed on a wire. Some are oriented moving to the right, and the rest to the left. When an ant reaches the end of the wire, it falls to the ground. When two ants meet, they both switch directions immediately. All ants move at the same constant speed, and it would take an ant one minute to walk the length of the wire had there been no other ants on the way to interfere.

Question. *How much time at most would it take for all the ants to fall off the wire?*

. . . .

Puzzle 5.2 **The Fast Ant**

Consider the initial setup of the "Ants on a Wire" puzzle, but where one of the ants drank a tiny drop of coffee, which made it walk just slightly faster than its 99 friends. As before, if two ants meet while moving toward each other, they both immediately reverse direction and keep their original speeds. If the faster ant catches up to a slower ant from behind, only the faster ant reverses direction; the slower ant continues forward uninterrupted.

Question. *How does the answer to the "Ants on a Wire" puzzle change? Would it take more or less time for all the ants to fall off the wire? Is the change significant or negligible?*

. . . .

Puzzle 5.3 **Ants on a Circle**

Now, the ants are randomly placed on the circumference of a circle, on which they are bound to walk. Some start walking clockwise, and others anti-clockwise. As before, when two ants meet, they change direction immediately. They all walk at the same speed, completing a full lap around the circle in one minute if it were empty.

Question. *How much time at most would it take for 99 ants to return to their initial position simultaneously? How long will it take for 100 ants?*

Puzzle 5.4 **Ants on a Square**

Four ants are placed, each on a different corner of a square. They start moving clockwise; each ant continuously follows the ant ahead of it. They all walk with the same velocity.

Question. *What is the distance that each ant would travel until they all meet in the center of the square?*

Puzzle 5.5 **The Drunk Ant Walk**

One ant walks on a stick. It takes identical tiny steps, and with each step it makes, the ant randomly decides whether to walk to the right or the left with a probability of 50% to each side. When the ant reaches the edge of the stick, it falls off. The stick is three meters long, and the ant is placed one meter from the right edge of the stick.

Question. *What is the probability that the ant will fall from the right edge of the stick?*

Puzzle 5.6 **The Drunk Ant Race**

Two drunk ants, A and B, are placed on the positive side of the number line. They both take identical tiny steps, and with each step they make, they independently decide whether to walk to the right or the left with a probability of 50% to each side and then take a small step in that direction. If they cross paths, they simply walk past each other. The first ant to reach position 0 wins.

Question. *Given that the probability of ant A winning is 90%, what is the probability that the ants won't cross paths?*

Curiosity. A Tourist Ant on a Box

Consider an ant living on a 1-by-2-by-3 box. One day, it decided to do a one-way tour to another point on the box, taking the minimal path. What is the maximum number of faces the ant could visit? Is it 3? Is it 4? No! It's 5! Here is an example:

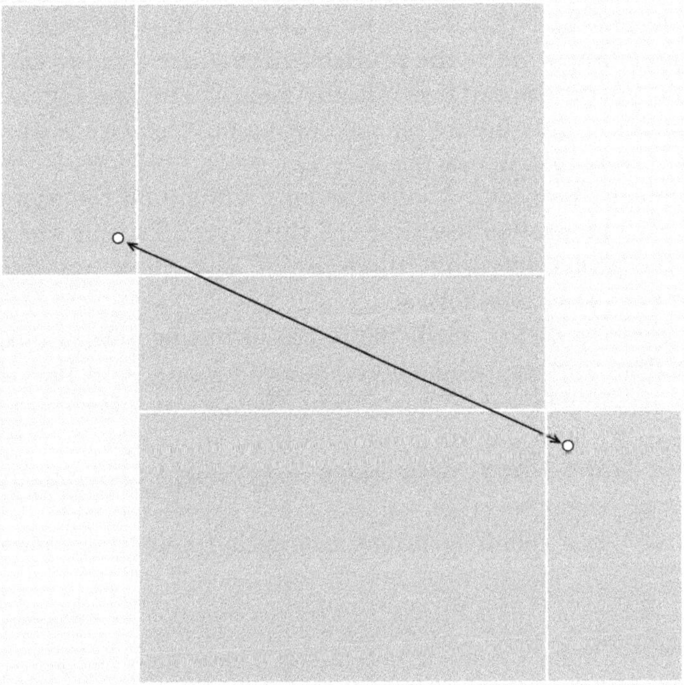

Source: The oldest reference to this problem we could find was https://hermann-sw.github.io/cuboid_shortest_paths/, although the problem seems to be much older, and it is likely that previous unreported solutions might have been known.

Solutions

Peter Winkler is a famous puzzle collector who wrote three books of puzzles. In his book *Mathematical Mind-Benders*, there is a separate chapter titled *The Adventures of Ant Alice*, which contains several ant puzzles.

It starts with a puzzle congruent to our first puzzle; then, Peter expands on it. For example, if ants were assigned the directions they are facing randomly, what is the probability that the third-from-the-left ant would fall off the end it is initially facing? The first big idea is that the number of ants falling off the east end of the wire is exactly the number of ants initially facing left. The second big idea is that if k ants face left, then exactly k leftmost ants will fall off the east end of the wire. So, the probability that the third ant falls from the east end equals the probability that there are at least three ants facing east. The final calculations follow.

Just like Peter Winkler, mathematicians like using ants in puzzles to represent dots, but we think there is more to ants than just their size. We are also impressed by how strong the ants are. For example, they can carry 10–50 times their body weight. Some research claims that they can support 5000 times their body weight. This means it is very easy for them to crawl on the walls or even upside down! No wonder we place them on boxes and walls to see them moving around.

Solution to Puzzle 5.1 **Ants on a Wire**

Sometimes, a puzzle that might sound very difficult to solve has the simplest solution. This puzzle is a beautiful example of that. With 100 ants, there is going to be an enormous number of collisions, and calculating the maximal distance (or, equivalently, the time) that an ant could walk may be challenging, even for gifted mathematicians. However, the solution doesn't involve complicated calculations at all, but rather only a simple realization. So, if you haven't solved the puzzle yet, you may want to stop reading here and try to do so.

The key observation of the solution is that ants are identical. Therefore, we may replace the formulation: "When two ants meet, they both switch directions immediately." with "When two ants meet, they continue walking undisturbed." In both cases, we have two ants that meet at one point and two ants that move away from that point. The identity of the ants doesn't matter for the puzzle. The second formulation is much easier to solve: The maximal time that an ant may stay on the wire happens if it starts on one end of the wire and walks towards the opposite end. This would take a mere minute, which is the answer to the puzzle.

Source: The puzzle appeared in Winkler's '*Mind-Benders*' as "Falling Alice." Before that, in the Emissary column, which references "We saw this on Ravi Shankar's puzzle mailing at the University of Oklahoma, and he found it on Francis Su's Math Fun Facts website at Harvey Mudd College" at https://legacy.slmath.org/attachments/media/news/emissary/EmissaryFall03.pdf.

Solution to Puzzle 5.2 **The Fast Ant**

Surprisingly, with a single slightly faster ant, the ants may take much more time to fall off the wire! This is why: Consider the case where the fast ant starts at one side of the wire and marches to the other side. In front of it, there is a slower ant that marches in the same direction. The faster ant may collide with the slower one right before it is about to fall, then it will change its direction towards the point where it started (this situation is impossible if all the ants have the same velocity). Assuming it will not meet another ant on the way back, the whole journey of that ant may take almost 2 minutes! Twice as much as before! Ironically, in this example, the ant that stayed the longest on the wire is the fastest. You may interpret the moral of the story as you wish.

Bonus. Can you find a configuration with one fast ant that will take more than two minutes for the ants to fall?

Source: This puzzle was developed by the authors in 2024.

Solution to Puzzle 5.3 Ants on a Circle

This one is trickier. We first notice that the ants maintain their order. This means, for example, that if ant B starts between ants A and C, it will maintain this order after any number of collisions. Then we notice that for a similar reason as in "Ants on a Wire", after every full minute, the ants would reach the same positions as they started with, but the individual ants themselves might rotate. For example, with three ants, after one minute, ant A could be in the initial position of ant C, which would be in the initial position of ant B, which would be in the initial position of ant A. In this example, it would take three minutes for all the ants to return to the same position, but how can we generalize it? We provide two solutions: a "mathematical" solution and a "physical" solution.

The mathematical solution:
Imagine that each ant is painted with a different color while wearing a hat of the corresponding color. Each time they collide, they swap hats. Let's follow the path of one hat. Note that:

1. The hats never change direction. That means if, for example, a clockwise-walking ant dresses in a blue hat, this blue hat will always rotate clockwise, even after collisions.
2. After one minute, each hat changes hands a number of times equal to twice as many ants that are walking in the opposite direction. To see this, visualize two hats going in opposite directions; they cross paths twice in a minute.

If we wish to find out when the ants will all be in their original position, then it suffices to find at which multiple of a minute they will have received their original hats back. If the total number of ants is N, and the number of clockwise-walking ants on the circle is C, then a hat goes back to its original owner in K minutes when $2C + 2C + \cdots + 2C$ (added K times) is a multiple of N. The first time this happens is when $2CK$ is the least common multiple of $2C$ and N. For odd N, we find that K can be as large as N, and for even N, we find that K can be as large as $\frac{N}{2}$. Therefore, for 99 ants, it may take up to 99 minutes for all of them to come back to their initial positions, and for 100 ants, surprisingly, it takes at most 50 minutes. That means the time was cut by half by adding an ant!

Comment: The final argument counts the number of clockwise walking ants C. It might seem as if the symmetry breaks between clockwise-walking ants and counter-clockwise-walking ants. However, we could have counted the number of counter-clockwise-walking ants W instead, and would have obtained the same result. Indeed, we have $C + W = N$, the least common multiple of $2C$ and N is the same as the least common multiple of $2W = 2N - 2C$ and N.

The physical solution:
Let us consider the difference between the number of ants that walk clockwise and the number of ants that walk counterclockwise D. This quantity is preserved after each collision and is therefore preserved all the time. In a minute, as a group, they will rotate by $\frac{D}{N}$, assuming the total number of ants is N.

If D in absolute value equals the number of ants, they will all rotate together and, therefore, reach their initial position after one minute. If $D = 0$, the ants, on average, would not rotate, and therefore, again, each ant would reach its initial position after one minute. If we add one ant to the latter case, making $D = 1$, the ants as a group would have a small rotation orientation on average, namely $\frac{1}{N}$. That means it would take the ants N minutes to reach their initial position together. This scenario is possible only for an odd number of ants. With an even number of ants, the smallest average rotation velocity can be $\frac{2}{N}$, and thus, it would take at most $\frac{N}{2}$ minutes for all the ants to return to their initial position.

Curiosity. **Ants and GCD**

Isn't it strange that odd and even numbers of ants behave so differently? In this curiosity, we give you the reason why: There are a few arrangements that require the same amount of time to reach the initial configuration. If we take D as in the solution above, it will take the ants $\frac{N}{\gcd(D,N)}$ minutes to return to their initial position, where gcd stands for the greatest common divisor. For an odd number of ants, D can be as little as one, and therefore, $\gcd(D,N)$ can also be as little as one. For an even number of ants, D can be as little as two (and always even), and therefore, $\gcd(D,N)$ can also be as little as two.

Source: This puzzle was developed by the authors in 2024. Peter Winkler has a question with the same name as ours in his book *Mathematical Puzzles* involving probabilities, his version asks instead "What is the probability that after one minute, every ant finds itself exactly where it began?".

Solution to Puzzle 5.4 — Ants on a Square

The initial configuration of the ants exhibits a symmetry under rotations by 90°. It follows that at each point in time, the ants form a square, and, at the end, they meet at the center of the square. The velocity of each ant is perpendicular to the velocity of the ant that it chases. That means the ant that is being chased doesn't cause the distance between them to expand or shrink. Its velocity doesn't have a component along the axis that connects it to the ant that chases it. Therefore, the sides of the square shrink at the same rate as if an ant were walking straight to the ant ahead of it. This means any ant would travel a distance equivalent to the side of the initial square.

Bonus. How would the answer change if the ants were initially placed on the corners of an equilateral triangle? Or, more generally, on the corners of an equilateral polygon? This may require a bit of trigonometry (but not too much!).

Source: Folklore.

Solution to Puzzle 5.5 **The Drunk Ant Walk**

The probability that the ant moves one meter to the right before it has moved one meter to the left is 50%. If it moves one meter first to the right, then it falls off, and if it moves one meter first to the left, then it is in a symmetric situation as it started, and the probability that it falls off either end inverts. Representing the desired probability by P, this gives the following equation:

$$P = \frac{1}{2} + \frac{1}{2}(1 - P)\,. \tag{5.1}$$

After simplifying this equation, we find that $P = 2/3$.

Source: Folklore.

Solution to Puzzle 5.6 **The Drunk Ant Race**

If the probability of ant A winning is 90%, then the probability of ant B winning must be 10%. This means that ant A must be closer to 0 than ant B. By how much? We don't know ... But what we do know is that for ant B to win, it must cross paths with ant A. And, if the paths intersect, then at the moment of intersection, we know that ant A and ant B must have an equal probability of winning. That means that the probability of ant B winning is equal to the probability of crossing paths times one-half. This implies that the probability of crossing paths must be 20%, and therefore, the probability of not crossing paths is 80%. This is the answer.

Source: This puzzle was developed by the authors in 2024.

Chapter 6

On Prophets and Sages

Socrates, the great Greek sage, once said: "All I know is that I know nothing". Isn't this a strange thing to say? If he truly meant what he said, put in other words, it must mean that he "knows nothing *except for* the fact that he knows that he knows nothing". However, isn't that admission in itself the knowledge of yet one more thing? Meaning he knows that he "knows nothing except for the fact that he knows nothing". However unconventional this correction might be, we propose a somewhat convoluted correction to the words of the sage: "All I know is that I know that I know nothing". Notice, however, that this unconventional correction fails to address what he now expresses! We are therefore forced to a yet more convoluted correction to the words of the sage, and from there, yet another, and so on and so forth, ad infinitum! Meaning, shouldn't the words of the sage be, "All I know is that I know that I know ..." and so on and so forth ad infinitum in a never-ending sentence expressing the long-awaited conclusion "... that I know nothing". Wise words often lead us to long and difficult paths! Who knows what Socrates meant? Did he *know*?

This chapter is about prophecies and riddles, two things that have gone hand in hand since the times of the prophet Ezekiel, whose famous opening words are: "Son of man, pose a riddle, and speak a parable to the house of Israel". Here, we will embark on a short, challenging, but meaningful journey through the words of prophets and sages. Can prophets prophesy that their prophecy will fail? Can the disagreement of sages be the key to interpreting hidden meanings? Can the knowledge of not knowing (like our beloved Socrates' Paradox explained earlier) be a key to the knowledge of knowing? We hope to find out in these meandering paths that await us.

Puzzle 6.1 **Random Number Prophecy**

It was once said that prophecy was given to the fools. Whether it is right or wrong, we are going to make a prophecy. Choose a **random** number between 1 and 100. Our prophesied number appears in the solution section.

Puzzle 6.2 **Two Random Oracles**

Before you are two oracles, both of which will answer yes or no to any question you ask them. One of the oracles will answer yes or no at random to any question you ask. The other, before each question, will first randomly decide whether to lie or tell the truth and then give the answer accordingly.

Question. *Can you tell them apart? If so, how?*

Puzzle 6.3 **The 77 Oracles — Part 1**

In a distant kingdom, some people are said to have acquired special powers. Calling themselves oracles, they claim to have the ability to enter the spirit realm and communicate with each other. The king wanted to test once and for all whether the oracles are truthful or charlatans.

The king set out to test 77 renowned oracles in the land. Without revealing who else or how many others were being tested, he asked them to delve into the spirit realm at midnight, shake hands with one another, and report back. After a long and sleepless night, they all reported having shaken hands with at least one oracle in the spirit realm; however, none of them could recognize who they saw. When the king compared their statements, he noticed something curious: each oracle claimed a different number of handshakes.

It is known that true oracles, if they exist, can travel to the spirit realm and never lie. False oracles are not capable of traveling to the spirit realm, so all their reports are bogus.

Question (1). *What is the maximum number of true oracles among the 77 who were tested?*

Question (2). *Assuming one oracle might have made a mistake by miscounting only by 1, how would your answer to Question 1 change?*

Curiosity. **Newcomb's Paradox**

In front of you are two boxes. Box A is transparent, and you see \$10,000 inside it. Box B is opaque. You have a choice: Take both boxes A and B, or take only box B.

Before you make this decision, you learn that a prophet prophesied in secret what your choice would be. The prophecy also comes with a peculiar setup: If he prophesied that you would take both boxes, he put nothing in box B. But if he prophesied that you would take only box B, he put \$1,000,000 in it. You are now in front of the boxes. What should you choose?

People are quite divided about what choice is correct. Choosing only box B seems logical as it would contain a larger prize, but by doing so, you seem to admit that your actions in the present have an effect on the past.

Puzzle 6.4 — The 77 Oracles — Part 2

As in the previous puzzle, assume that there are 77 oracles who are being tested. This time, all of them can access the spirit realm, but some oracles were less than honest (trying to boast their skills) and reported 1 more than the actual number of hands they shook. Yet again, the strange thing was that every oracle reported shaking hands with a different number of people.

Question. *The king wants to hire the most proficient oracle (the one who shook the most hands in the spirit realm) that gave an honest report. How many hands did that oracle shake?*

Puzzle 6.5 **Fitch Cheney Trick**

This one is a famous card trick that requires no sleight of hand. A magician lets a random crowd member choose five cards from a standard deck of 52 cards. The spectator shows the five cards to the assistant, who then chooses which four to reveal to the magician, and in what order. The magician then correctly predicts the fifth card.

Question. *The only information that the assistant gives to the magician is the four cards and their order. How could the magician always correctly guess the fifth card?*

Puzzle 6.6 **Tale of the Four Prophets**

A long time ago, four prophets spoke. Ever since, no more prophets have ever nor will ever come to be under the sun. Each prophet came to the land after the other had passed away, and each knew and agreed with their predecessors' prophecies: The second prophet knew and agreed with the first prophecy, the third prophet knew and agreed with the first two prophecies, and finally, the fourth prophet knew and agreed with the first three prophecies.

The sacred writings testify that one of the prophets said:

"*There will be or has been*
at least one ***false*** *prophecy*
uttered by a ***false*** *prophet*".

Another prophesied that:

"*There will be or has been*
at least one ***true*** *prophecy*
uttered by a ***true*** *prophet*".

Yet another prophesied:

> *"There will be or has been*
> *at least one* ***true*** *prophecy*
> *uttered by a* ***false*** *prophet".*

Finally, one prophesied:

> *"There will be or has been*
> *at least one* ***false*** *prophecy*
> *uttered by a* ***true*** *prophet".*

Each prophecy in the list above was prophesied by a different prophet. However, **the order in which these prophecies were uttered remains unknown to us**.

It is known that a true prophet only says true things and never agrees with false prophecies. It is also known that a false prophet may or may not prophesy something true; a false prophet is false either because he prophesied a false prophecy, or because he agreed with a false prophecy, or both.

Now it came to pass that three wise men, who knew the order in which the prophets came, each found a consistent way to interpret which prophecies were fulfilled and which weren't. When they compared their notes, they discovered that they all disagreed. Thankfully, in the first great council, chaired by the three wise men, after heated debates and long meditative moments of prayer, they finally decided to resolve the conflict by simple majority voting: they placed votes for each of the four prophecies simultaneously, deeming each one to be either true or false. To their surprise, the majority outcome of all of the votes completely aligned with the opinion of the eldest of the wise men, and from that moment onward, he was considered the greatest among equals.

Question. *In which order did these prophecies come? According to the council decision, which prophecies were true and which were false?*

Solutions

This chapter is reminiscent of many traditions of wisdom and theological teachings passed down through generations. Previous recreational mathematics authors have also written about the seemingly unlikely, yet also unavoidable, crossing of paths between mathematics and faith, between the exact sciences and the mystical arts.

The famous Martin Gardner, a beloved author in recreational mathematics, playfully wrote about the whimsical Dr. Matrix and his incredible numerological findings as recorded in many of his works [9]. Malba Tahan, yet another beloved recreational mathematics author, wrote much of the sober desert wanderer seeking to obey the Almighty while uncovering jaw-dropping mathematical gems [10]. Both of these authors, whose works equally influence the genesis of this chapter, capture the experience of faith and mathematics from different yet complementary angles: one more playful and the other focused and intentional.

Here, we hope that this chapter will inspire both young and experienced mathematical minds to never stop using their love of numbers, not only in the mundane matters of life but also in the profound and spiritual affairs that they might encounter. And, just as the giants that preceded us, to carry through with both playfulness and reverence.

Solution to Puzzle 6.1 Random Number Prophecy

Our prophecy is: The number that you chose is 37! Did we get it right? If so, you are probably amazed, but if not, don't be too disappointed that you got it wrong. It seems that when people are asked to choose a random number between 1 and 100, the most common answer, with a probability much greater than 1/100 (although the exact value is not clear), is 37. We encourage you to repeat the experiment with your family and friends and see the result.

Source: Check out http://thirty-seven.org. It collects random occurrences of 37, which appear not so random after all.

Solution to Puzzle 6.2 Two Random Oracles

The answer is... Yes! You can distinguish between the two random oracles. But how? The trick is to recognize that the oracle that randomly decides to tell the truth or to lie when given the question, "Are you answering this question truthfully?" would always answer "Yes". However, the other oracle may answer "No" to the same question. Therefore, repeat this question to both sides until you receive a "No", and you will tell them apart. It might take a long time...

Source: This puzzle was developed by the authors in 2024.

Solution to Puzzle 6.3 The 77 Oracles — Part 1

Solution to question 1:
This question seems impossible since we can't say whether an oracle saw a fraction of the other oracles in the spirit realm or simply lied about what they saw. Nor can we double-check the compatibility of the numbers they informed the king, or can we?

First, let us recognize that there can't be exactly one truthful oracle in the kingdom since a solitary oracle would be forced to see no other oracle in the spirit realm. Now, the trick lies in realizing that if two or more oracles were telling the truth, then at least two of the truthful oracles must have shaken hands with the same number of oracles in the spirit realm. To see this, pretend that, from the 77 oracles, a total of 13 were truthful oracles, and all of them claim to have shaken hands with at least one oracle. Each of these oracles can shake hands with at most 12 others... and therefore, each of the 13 oracles must be assigned a number from 1 to 12, forcing at least two of them to have the same number. Therefore, there can't be two or more truthful oracles in this kingdom either... Thus, the answer is: all the oracles in this kingdom must be lying.

Solution to question 2:
If one oracle miscounted in the spirit realm, then there can be at most two truthful oracles. To see this, assume, for example, that there was a subset of 13 truthful oracles. If they all said different numbers, then they must have said the numbers 1 to 13. The oracle who said 13 must have been mistaken by adding 1. Thus, before the mistake, they would have the numbers 1 to 11 and then twice the number 12. However, two people claiming to shake hands with 12 is incompatible with one person with the number 1.

The argument above doesn't depend on the number 13 but only on the fact that the subset of truthful oracles is greater than 2. For 2 people, we can easily verify that one miscount can produce the numbers 1 and 2.

Source: This puzzle was developed by the authors in 2024. It is a variation of the famous hand-shaking party puzzle.

Solution to Puzzle 6.4 — The 77 Oracles — Part 2

If all numbers are different, then the oracles must have reported the numbers 1 to 77. It is clear, however, that the oracle that reported

77 must have inflated his count since no one can shake hands with more than 76 oracles. This implies that the oracle that reported 1 must be telling the truth.

The true number of handshakes of an oracle that reported N handshakes will be represented with $T(N)$. Thus, above we have proven that $T(77) = 76$ and this implies that $T(1) = 1$.

Now, consider the oracle that reported the number 76. Such an oracle couldn't have shaken hands with the oracle who reported 1 and could have shaken hands with all oracles reporting from 2 to 75 as well as 77; that makes 75 in total. Therefore, $T(76)$ must be 75. And, similar to our previous conclusion, $T(76) = 75$ implies that $T(2) = 2$.

This reasoning can be continued in this fashion: we find that $T(75) = 74$, because there aren't enough oracles for $T(75) = 75$, and this implies that $T(3) = 3$. The process leads up to $T(40) = 39$ which implies that $T(38) = 38$. This leaves us with:

$$\begin{aligned} T\,(1) &= 1,\\ T\,(2) &= 2,\\ &\vdots\\ T\,(38) &= 38,\\ T\,(39) &= ?,\\ T\,(40) &= 39,\\ &\vdots\\ T\,(77) &= 76. \end{aligned}$$

We are left with the question of whether $T(39) = 39$ or 38. The answer is $T(39) = 38$ since that oracle must have shaken hands with oracles reporting 40 to 77 and can't have shaken hands with 1 to 39.

Therefore, among the honest oracles, the one who shook the most hands was the one who reported 38 handshakes.

Source: This puzzle was developed by the authors in 2024. It is a variation of the famous hand-shaking party puzzle.

Solution to Puzzle 6.5 — Fitch Cheney Trick

Before the trick is done, the magician and the assistant pre-arrange a secret code to translate the order of the first four cards into information that uniquely determines the 5th card. Here is how it is done:

Since the deck of cards contains four suits, and five cards were selected, it is guaranteed to find two cards of the same suit[1]. The assistant and the magician agree that the suit of the first card shown to the magician determines the suit of the fifth card. The assistant has a choice: which card out of the two with the same suit will be the first card, and which will be the fifth? We will get to it in a moment. For now, the magician can determine the suit of the fifth card from the first card, and the next three cards should determine the number of the fifth card. There are $3! = 6$ ways to permute the order of three cards. The magician and the assistant then also agree on the order of the cards in the deck. There are many ways to choose such an order. For example, they may choose that a higher card is ranked above a lower card, and the suit of the cards is a tie-breaker, say diamonds > clubs > hearts > spades. Using the chosen order, the three remaining cards in the assistant's hand are ordered first, second, and third. By the order in which they are shown to the magician, the assistant can communicate a number between 1 and 6. Here is an example of such ordering:

Cards order	**Number communicated**
1st, 2nd, 3rd	1
1st, 3rd, 2nd	2
2nd, 1st, 3rd	3
2nd, 3rd, 1st	4
3rd, 1st, 2nd	5
3rd, 2nd, 1st	6

[1]This is known as the pigeonhole principle: If n pigeons fit into $n-1$ pigeonholes, there would be at least one pigeonhole with at least two pigeons.

But wait, how can the magician determine the number of the fifth card if there are 13 possibilities and only 6 numbers that the assistant can deliver? Remember that the assistant has a choice of which of the same-suit cards to show the magician as the first card, and which to keep in hand as the fifth card. Note that exactly one of the two scenarios must occur regarding the numbers of the two same-suit cards: (1) The higher number is larger by 6 or less than the lower number. (2) The higher number is larger by 7 or more than the lower number. If the first scenario occurs, the lower number will be chosen as the first card. The magician will add to its value the number that is represented by the order of the remaining three cards. If the second scenario occurs, the higher card will be chosen as the first card. Also here, the magician will add to its value the number that is represented by the order of the remaining three cards; and if the number goes beyond 13, he subtracts 13.

For example, if the numbers on the two same-suit cards are 3 and 7, the first card will be 3, and the assistant will arrange the following three cards in an order that will deliver the number 4. The magician will then call the suit of the first card, and the number $3 + 4 = 7$, which is indeed the fifth card. If, for example, the numbers on the two same-suit cards are 2 and 10, the first card will be 10, and the assistant will order the remaining three cards in an order that will communicate the number 5. The magician will then call the suit of the first number, and $10 + 5 = 15$, but since this number is greater than 13, we subtract $15 - 13$ and find 2, which is indeed the fifth card.

Source: Wallace Lee, *Math Miracles*, published by Seeman Printery, Durham, N.C., 1950. Also, Tanya and her students discovered many generalizations of this trick and wrote about them in their paper: 'Card Tricks and Information' available at: https://arxiv.org/pdf/2405.21007.

Curiosity. A Bigger Deck of Cards

By the way, the strategy described in the solution only works for decks of sizes up to 52 cards. So, the standard deck is a perfect fit. However, there is a different algorithm that works for decks of sizes up to 124. Moreover, Michael Kleber, in the paper 'The best card trick' (*The Mathematical Intelligencer*, 24 (2002), 9–11), used information theory to prove that the maximum size is, in fact, 124.

Solution to Puzzle 6.6 Tale of the Four Prophets

For convenience, we label the four prophecies according to the table:

Prophecy **FF**	There will be or has been, at least one **false** prophecy uttered by a **false** prophet.
Prophecy **FT**	There will be or has been, at least one **false** prophecy uttered by a **true** prophet.
Prophecy **TF**	There will be or has been, at least one **true** prophecy uttered by a **false** prophet.
Prophecy **TT**	There will be or has been, at least one **true** prophecy uttered by a **true** prophet.

It is possible to immediately notice that prophecy **FT must be false**, as no true prophet would prophesy a false prophecy. This means that prophecy **FF must be true**, as prophecy FT is a false prophecy uttered by a false prophet.

Because the three sages interpret the nature of prophecies differently, they have to disagree about at least two prophecies. Thus, the truthfulness of prophecies TT and TF must remain ambiguous. We also notice the following:

1. If prophecy FF came after FT, that makes the prophet of FF a false prophet (because he agreed with a false prophecy that was uttered before him), which means prophecy TF must be true. That wouldn't leave room for all three sages to disagree, and we thus conclude that **prophecy FT came after prophecy FF** (but maybe not right after).
2. If FF was the first prophecy, its prophet must have been a true prophet, which means prophecy TT must be true. That wouldn't leave room for three sages to disagree, and thus prophecy **FF wasn't first**.
3. Suppose TF was the first prophecy. If the prophecy TF is true, then the first prophet is a true prophet, implying that TT is true. If TF is a false prophecy, then every one after is a false prophet, implying that TT is also false. As there are only two

cases, that wouldn't leave room for all three sages to disagree, and thus **prophecy TF wasn't first**.

Therefore, from 1, 2, and 3, we conclude that **prophecy TT was the first** under the sun. Now, prophecy TF could be 2nd, 3rd, or 4th. Let's explore these options:

- Suppose TF came 2nd under the sun, implying that the order is TT, TF, FF, FT.

  If TT is false, then FF must be a true prophecy by a false prophet, implying that TF is true. If TT and TF are true, then the first three prophecies were uttered by true prophets. Prophecy FT is, as always, false, which means there was no true prophecy by a false prophet, in contradiction to the assumption that TF is true. Thus, if TT is true in that considered order, TF must be false.

  It follows that the truthfulness of TT is opposite to that of TT, which wouldn't leave room for three sages to disagree.
- Suppose TF came 3rd under the sun, implying that the order is TT, FF, TF, FT.

  If TT were true, there would be a paradox between prophecy TF and itself (If TT is a true prophet, then so is FF. That means, for TF to be false, it must be uttered by a true prophet, which is a contradiction, and for it to be true, it must be uttered by a false prophet; but since there was no false prophet before it, it also can't be true), which means TT must be false and the sages would agree on it unanimously.

These facts uniquely determine the position of Prophecy TF to be the fourth, and thus, the full order is: **TT $\rightarrow$ FF $\rightarrow$ FT $\rightarrow$ TF**.

Prophecies TT/TF could be true/true, true/false, or false/true (but not false/false, as if TT is a false prophet, so do the other prophets, which makes FF a true prophecy, which makes TF a true prophecy). After the sages vote for these options, we conclude that prophecies TT, TF, and FF were fulfilled, while prophecy FT wasn't.

Source: This puzzle was first published in the G4G15 Book exchange under the name *The Four Prophets and the Three Sages* by Ivo David and Yogev Shpilman in 2024.

Chapter 7

Can You Recognize the Pattern?

[Ancay Ouyay Ecognizeray ethay Atternpay?]

People love recognizing patterns. This inclination is called patternicity, and it helps humans make sense of the world. We love looking for patterns so much that we sometimes see connections where there are none! There is also a special term for this phenomenon: it is called apophenia.

What is strange, though, is that who is to say that a pattern is there or not? For example, say one day we detected blinking lights coming from the surface of a distant planet, and they encoded some kind of Morse code sequence "011101001...01" and then, suddenly, the blinking lights stopped! Wouldn't the world go crazy trying to find out if we might have received a message from a distant alien civilization? What if it is all random? Could we tell the difference?

Somehow, AI has found patterns in how we speak and the text we produce... and it seems to use these patterns to strangely talk back to us, almost as if the pattern were really there... Could AI crack the patterns of a message from a distant civilization? And if it did, how could we know it was right?

In this chapter, you will get a chance to feed your patternicity itch. Find the missing element, detect the odd one out, and find the rule guiding the sequence.

Curiosity. **A Pattern You Might Not Know**
In order to inspire instincts of patternicity for this chapter, we provide an illustration of what seems like an alien language. It is a page from a notebook of the artist Raoul-Auger Feuillet, dated to 1701. Before reading any further, can you guess what the patterns in the drawing mean?

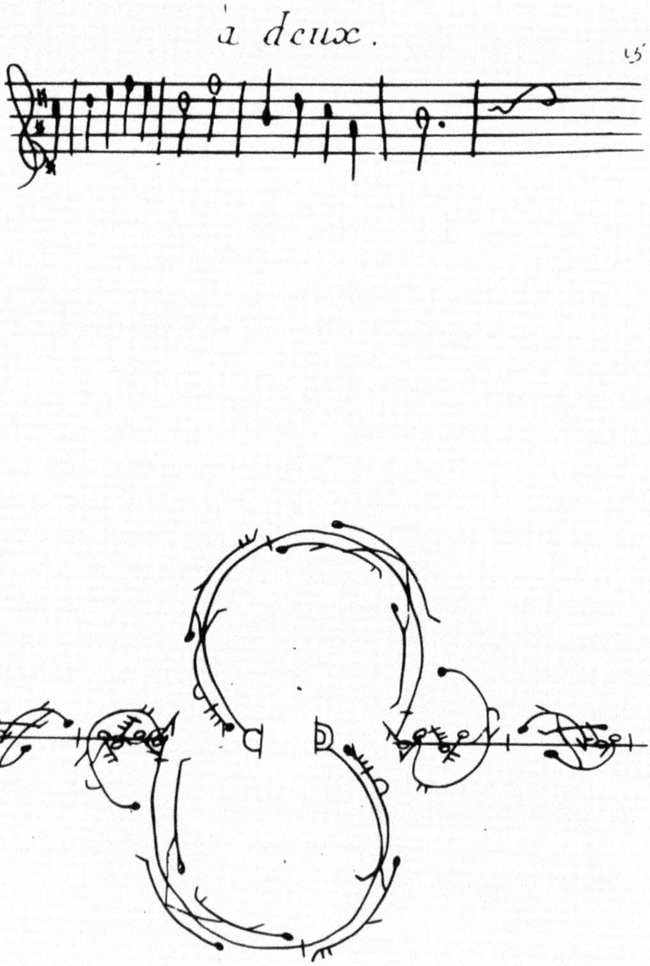

This is one amongst many illustrations of a language that encodes the fluid movement of dance alongside the music it is danced to. Today, it is referred to as the Beauchamp–Feuillet notation, an early dance notation system believed to have been invented in the 1680s. Unlike in the next patterns you are about to see throughout the chapter, here, recognizing the pattern means identifying how the body must move in order to follow the notation. Can you recognize the pattern?

Puzzle 7.1 **Find the Odd One Out**

Consider the shapes as follows.

`the Owl sequence`

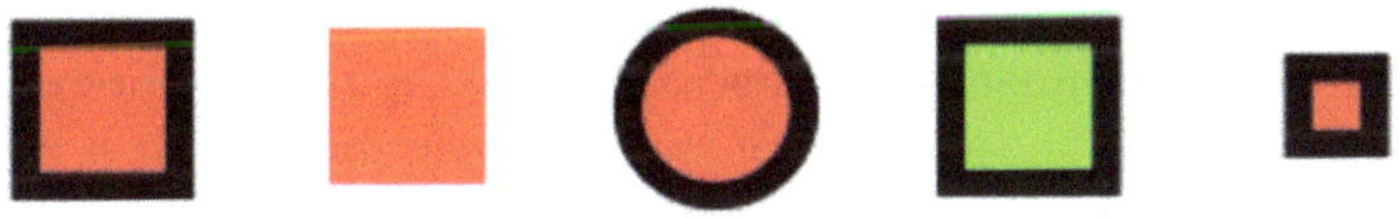

Question. *Which shape is the odd one out?*

Puzzle 7.2 **What Comes Next?**

Consider the sequence as follows.

`the Robin sequence`

`O, T, R, F, I, S, ?`

Question. *Which letter should replace the question mark?*

Curiosity. **The Sausage Catastrophe**

Mathematicians have long studied how to neatly arrange objects in a contained region. The sausage catastrophe is an odd situation that happens when you ask the question: if you were to pack n oranges in a convex package[a] what is the smallest volume for the package?

In 1992, the mathematicians Jörg Wills and Pier Mario Gandini conjectured that the optimal packaging for $n = 1$ up to 55 is the sausage package and that the pattern breaks at $n = 56$. In fact, the sausage package pattern breaks for $n = 56$, 59, 60, 61, 62, and for all $n \geq 65$, what a catastrophe!

[a] A convex region is one where any two points can be connected with a line segment, and the line segment will still belong to the region, e.g., a square ■ is convex but the letter **U** is not.

Puzzle 7.3 **What's Missing?**

Consider the shapes as follows.

`the Duck sequence`

Question. *Which shape is missing?*

Puzzle 7.4 **Replace Question Marks**

Consider the following sequence of **Y**s and **R**s.

`the Eagle sequence`

`Y, Y, ?, ?, Y, ?, Y, ?, R, R, R, R.`

Question. *Which letters should replace the question marks?*

Puzzle 7.5 **Simplify the Expression**

Consider the expression:

`the Raven expression`

$$\ln\ln \mathbf{a} \times \ln\ln \mathbf{b} \times \cdots \times \ln\ln \mathbf{z}$$

where ln is the natural logarithm.

Question. *Can you simplify it any further?*

Curiosity. **Moser's Circle Problem**

Draw n points on the circumference of a circle, and then draw every possible string between the points. How many areas did you divide the circle into? For $n = 2$, 3, 4, and 5, we get that the number of areas is 2, 4, 8, and 16. The pattern is clear: the number of areas as a function of the number of points is 2^{n-1}. So, for $n = 6$, we should get 32 areas. However, if you count carefully, the actual number is 31. Such a small mismatch, and yet, the pattern is broken!

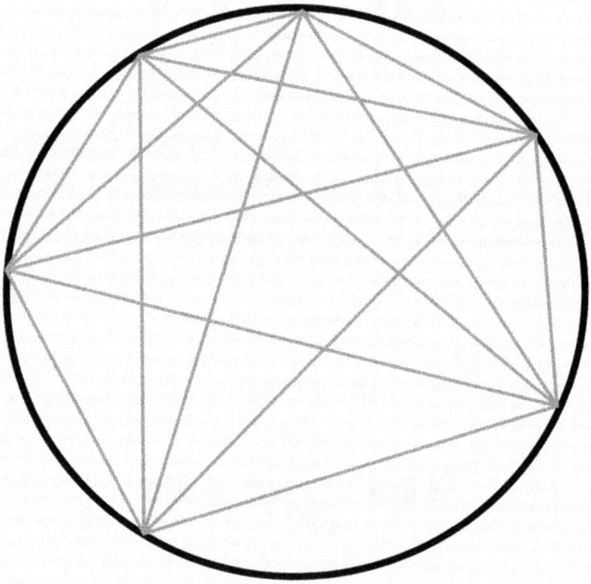

Solutions

How would you continue the sequence 1, 2, 3, 4, 5, 6, 7, 8, 9? Of course, the next number must be 10, right? What if we told you that for some *cool people*, the next number is 153? How would you feel about that? Yes, that is the answer, and before you get angry, we can explain!

The pattern we are thinking of is called *the narcissistic numbers*, where the next number is 153. Here is how it works: The sequence contains all natural numbers that satisfy the following property: sum all the digits that form the number to the power of m, where m is the total number of digits. If the sum produces the original number, then we say the number is narcissistic. For example, take the number 83; it has two digits, does $8^2 + 3^2 = 83$? No. So, 83 is not part of this sequence. On the other hand, every 1-digit number is narcissistic. For example, it is easy to check that $8 = 8^1$. So, it's kind of strange that none of the narcissistic numbers have two digits. The next one after 9 is 153 (one can check that $1^3 + 5^3 + 3^3 = 1 + 125 + 27 = 153$), and then 370, 371, 407, and so the sequence continues.

As you can see, the most obvious choice (of 10) is not always the right answer. Hopefully, you will get most, if not all, the answers in this chapter correct. But despair not if you don't! Maybe you just found the *cooler* pattern that we couldn't find.

PS: As of April 2025, AI has not been able to solve even one of the puzzles of this section! If you solve even just one, you can give yourself a pat on the back for beating AI in its own game: pattern recognition.

Solution to Puzzle 7.1 — Find the Odd One Out

The second shape is the odd one out as it is the only one without a border. The third shape is the only one that is not a square. The fourth shape is the only one that is green; the last shape is the only one that is small. Thus, all shapes but the first are odd ones out in some sense. **Which makes the first one the odd one out.** But wait, now all of them seem to be the odd ones out. Hmm.

Source: This is original by Tanya Khovanova, invented in 2009.

Solution to Puzzle 7.2 — What Comes Next?

The first seven numbers share the following property: they each share one letter with the number ahead of it:

1. One & Two share an O,
2. Two & Three share a T,
3. Three & Four share an R,
4. Four & Five share an F,
5. Five & Six share an I,
6. Six & Seven share an S.

The shared letter is our sequence. Thus, for (7), we have that Seven & Eight share an E. Therefore, the answer is E.

Source: This puzzle was developed by the authors in 2024.

Solution to Puzzle 7.3 — What's Missing?

We can assume that there are three reds and three greens, three squares and three circles, three small shapes and three large shapes, three shapes with borders and three without. Thus, the missing shape is a small red square without a border.

Source: This is original by Tanya Khovanova and Gregory Bomash from 2010.

Solution to Puzzle 7.4 **Replace Question Marks**

These are the last letters of the months in order. The answer is:

Y, Y, H, L, Y, E, Y, T, R, R, R, R

Source: This is based on a well-known puzzle. The oldest variant we could find, though older versions almost certainly exist, is from the *Giant Book of IQ Puzzles* by David J. Bodycombe, first published by Constable & Robinson Ltd. in 1997.

Curiosity. The Legend of the Ten-Month Calendar
Have you ever noticed that *December* is named after the Latin word "decem" (ten) even though it is the 12th month of the year? The same goes for *September*, *October* and *November*, meaning "septem" (seven), "octo" (eight) and "novem" (nine).

This is because, in earlier Roman calendars, the year began in March. Since winter wasn't relevant for agriculture, the months of January and February weren't part of the year. In those times, the calendar would only account for 304 days of the year, leaving the rest out of reach of the counting system. And, today, what is left are these curious linguistic patterns, a reminder of a long-gone ten-month year.

Solution to Puzzle 7.5 **Simplify the Expression**

Since $\ln e = 1$, we have $\ln \ln e = 0$, and the whole expression is simply zero!

Source: This is the puzzle by Tanya's son Alexey Radul.

Chapter 8

ImProbabilities

To kick off the mood for what you are about to encounter in this chapter, consider the classic St. Petersburg Paradox: The story goes that you are offered a ticket to play a game where a fair coin is tossed until it lands heads, and the payoff doubles with every toss: \$2 if it lands heads on the first try, \$4 on the second, \$8 on the third, and so on. The expected value of this ticket is

$$2 \cdot \tfrac{1}{2} + 4 \cdot \tfrac{1}{4} + 8 \cdot \tfrac{1}{8} + \cdots = 1 + 1 + 1 + \cdots = \infty.$$

But would you pay an infinite number of dollars for a one-shot chance to play this game? How about \$1000 or \$10?

Now, meet its lesser-known cousin: the Pasadena game[1]. It's just like the St. Petersburg game, but with a twist: The payouts go up and down in alternating fashion, taking the values positive 2, negative 4, positive 8, and so on. The strange result? The expected value of this game isn't just huge or tiny. It's undefined! Not infinite, not zero, just mathematically... off:

$$2 \cdot \tfrac{1}{2} - 4 \cdot \tfrac{1}{4} + 8 \cdot \tfrac{1}{8} - \cdots = 1 - 1 + 1 - 1 + 1 - 1 + \cdots = ?$$

[1] Actually, we're showing here a simplification of the original Pasadena game. For the original, more nuanced version, refer to [13].

So, what would you pay for a ticket to a game that doesn't even know what it's worth?

In this chapter, we won't ask you to compute infinite series or debate convergence, but we will challenge your gut feelings about what's probable, what's fair, and what makes sense. You'll find puzzles where rare events pop up suspiciously often, where averages lie through omission, and where the most intuitive answer is often the least probable one.

Before we get into it, though, let's get one thing straight: Probabilities and improbabilities alike depend heavily on modeling! Just like in the *Patterns* chapter, you can expect some interesting problems that might just require some heavy lifting on your side; i.e., coming up with some really clever modeling assumptions; i.e., things that are not quite fully stated, but that just make sense given the context. Some of the problems don't have much of this at all, and yet others require a lot of it! "Which ones?" I hear you ask... well... if you really want to know in advance, then our polynomial $m(x)$ is the key:

$$m(x) = x^5 - 34x^4 + 441x^3 - 2684x^2 + 7460x - 7200.$$

It equals zero for each problem number x that requires modeling. Though we encourage you to try out each problem as it comes, without checking.

Puzzle 8.1 **Chess Tournament**

In a chess tournament, to win a prize, you need to win two consecutive games out of three. Assume that there is a higher probability that you beat an amateur than a master.

Question. *Would you prefer playing against an amateur $\rightarrow$ master $\rightarrow$ amateur, or master $\rightarrow$ amateur $\rightarrow$ master?*

Puzzle 8.2 **February 29**

Pick a random day in February.

Question. *What is the probability that the day is February 29?*

Puzzle 8.3 **Can't Beat Nature?**

In a distant village, every couple continues having children until they give birth to a boy. Once they have a boy, they stop having kids.

Assume all families have exactly one boy at the time of counting for the following questions.

Question (1). *What is the expected ratio of boys to kids in the population of that village?*

Question (2). *How does the answer to Question 1 compare to the expected value of the ratio of boys to kids in each family in that village?*

Puzzle 8.4 **This Is Unfair!**

You are given an unfair coin with an unknown probability of landing heads or tails.

Question. *Can you design a game that simulates a fair coin using this coin?*

Curiosity. **Coprime**

Two natural numbers are said to be coprime if the only natural number that divides both of them is 1. In other words, if the fraction $\frac{a}{b}$ can't be further reduced, then a and b are coprime.

One may wonder, what is the probability that two random numbers are coprime? What do we mean by a random number? If we take randomly two numbers below N, and then allow N to tend to infinity, then we get a quite beautiful answer, and it is exactly $\frac{6}{\pi^2}$. Where did π come from?!

Puzzle 8.5 **Tuesday Glitch**

You overheard disconnected parts of a conversation, where a man speaking on the phone said "I have two kids, one is a girl" and later the same man said, "My daughter stole a cake on Tuesday". You heard nothing more as the man passed by.

Question (1). *Before hearing of the cake thief, what is the probability that the man has two daughters?*

Question (2). *After hearing of the cake thief, what is the probability that the man has two daughters?*

Puzzle 8.6 **Math Department Trips**

Every day, 100 math professors commute to the department. There are 10 buses available, and each professor randomly chooses a bus for the ride. On average, each bus takes 10 professors. However, the professors all report that on average, there are more than 10 professors on their bus.

Question. *Can you resolve the paradox? How is this possible?*

Curiosity. **A Coin Toss Is Not Quite Fair**

The coin toss is not exactly 50–50. Experiments show that it's actually more like 51–49! That is, the coin more often lands on the same side it started with about 2% advantage.

Source: In 2023, a group of crazy statisticians published this strange finding in a paper called "Fair Coins Tend to Land on the Same Side They Started: Evidence from 350,757 Flips" [2].

Puzzle 8.7 **Battleships**

As a general in the Navy, you must choose one of the options:

- Sending on a mission one big ship, which has a probability p of success.
- Sending on a mission two small ships, each has a probability $\frac{p}{2}$ of success. At least one of the ships needs to be successful for the mission to be completed.

Question. *Which of the options would you take to maximize your chance of success?*

Puzzle 8.8 **Missing 5**

Alice rolls a die until she gets 6. Then Bob observes that she never rolled a 5.

Question. *What is the expected number of times that Alice rolled the die?*

Puzzle 8.9 **100 Prisoners, 1 Bullet**

You are a guard in a field with 100 prisoners, and your goal is to keep them from running away. Each prisoner will try to escape if they think they have a non-zero probability of success. The prisoners know that you have only one bullet in your gun, and you may make an announcement to share with the prisoners what you would do if anyone tries to escape. After they hear your announcement, the prisoners may strategize and cooperate with each other. Both you and the prisoners know you are a perfect shooter.

Question. *Can you design a shooting strategy that will prevent the prisoners from escaping?*

Puzzle 8.10 **Birthday's Bet**

Tanya was invited to give a probability lecture at a public school. She explained the famous birthday paradox to the students, which states that the probability that among 23 random people, two people share a birthday is close to $1/2$. Tanya noticed that there are 22 students in the class, and mentioned that the probability of having a shared birthday among 22 random people is slightly less than $1/2$. To engage her students more, Tanya decided to bet that there are no shared birthdays in the class.

Question. *Was Tanya's bet a good idea?*

Solutions

To close this chapter, let's play a little game. Before we started writing this introduction, we flipped a fair coin. If it landed heads, we promised ourselves we would *definitely* hide a secret message somewhere in the text. If it landed tails, we left it to whim, we'd only hide something if we truly felt inspired to do so.

Before you start searching the text, what is the probability that the coin landed tails? Easy, right? Just a 50/50 split. But what about after you've completed the search?

Here's where things get interesting: after you find a secret message, or if you find none, the odds seem to change in more than just one way. The number might change, depending on how confident you are in your code cracking skills, and the meaning of the number might also change, just so in a strangely beautiful and subtle way.

Think about it, before you search, the probability seemed to refer to "something out there", the cold mechanics of frequencies on coins and tosses. And after you search, the probability seems to refer to "something in you", a reflection of how confident you are in the evidence. On one side, expressing raw frequencies, on the other, measuring confidence.

Isn't this beautiful? When we say a probability is 50%, it might mean "I have no clue" or it might mean the much more precise "I'm pretty sure there is a random process where the outcomes occur with equal frequency". The same number, two very different ideas. That's the quirky, beautiful ambiguity of probability, something you won't easily find in other branches of mathematics.

If you are interested in finding out more about this dual nature of probability, then you might consider searching for: "epistemic vs aleatoric uncertainty". It's quite fascinating to realize this fluid and unusual topic in the mathematical sciences. And, there are actually other types of probabilities... but that is a deep and dangerous rabbit hole, we warn you, and if you search for "frequentist vs Bayesian probabilities", we quite honestly don't know how astray you might go.

Curiosity. **Weather Forecast**

Every day, John goes home on foot. Depending on the weather, he takes the path through the park, city, or coastline with different probabilities:

Path through the:	**Park**	**City**	**Coast**
Sunny Day	50%	30%	20%
Cloudy Day	40%	40%	20%
Rainy Day	10%	60%	30%

Given that John took a path through the park last Monday, what is the probability that it rained?

Lesson. One very common mistake is to say 10%. But why is this wrong? The key lies in understanding the difference between *probability* and *likelihood.*

Probabilities always add up to 100%. Note that this is the case for any given weather: the probabilities in the corresponding row add up to 100%. However, the columns add up to 100%, 130%, and 70%, respectively; how is this possible? Each column makes up what is called a likelihood function, not a probability! What is the likelihood, you ask? It is a measure of how much a given event supports a hypothesis: Walking through the park corresponds to a 10% likelihood that it was a rainy day; however, nothing can be said about the probability.

There you have it. A tiny little probability vs likelihood lesson before we get further into some other puzzles. Welcome to the club! And, don't let them foolya next time you see the weather forecast saying "a probability of X% that...".

Solution to Puzzle 8.1 — Chess Tournament

Your primal instinct may tell you to avoid two matches with the master. However, in order to win the prize, you must beat him at least once. Therefore, in order to maximize your probability of winning the prize, you should confront the master twice and choose master → amateur → master. This is true as long as the probability of beating the amateur is larger than the probability of beating the master.

Explicitly, if we let p_1 be the probability of beating the amateur, and p_2 the probability of beating the master, we have that the probability of winning the prize while competing against amateur → master → amateur is

$$p_2\left(1-(1-p_1)^2\right).$$

The p_2 term is here because you must win the middle game to get the prize, and the $1-(1-p_1)^2$ term is the probability of not losing both the first and the third game. Similarly, the probability of winning a prize while competing against master → amateur → master is

$$p_1\left(1-(1-p_2)^2\right).$$

After expanding, we find that the difference between the two probabilities is

$$p_1p_2\left(p_2-p_1\right)$$

which is negative since $p_2 < p_1$. Therefore, for any values of p_1 and p_2, the option master → amateur → master is preferable as long as $p_2 < p_1$.

Source: The book *Fifty Challenging Problems in Probability* by Frederick Mosteller.

Solution to Puzzle 8.2 — February 29

This puzzle is ambiguous, as it is not clear which options a random day is chosen from.

Assuming one uses a Gregorian calendar, here is an approximate calculation. In a four-year cycle, there are $4 \cdot 28 + 1 = 113$ days in February. Thus, the probability is $\frac{1}{113} \approx 0.00885$. However, the leap years don't always repeat every 4 years. The rule is: Leap years occur every four years, except for centennial years (years ending in 00) that are not divisible by 400. Thus, the Gregorian calendar cycles through 400 years with 97 leap years. That means, in the full cycle, we have $400 \cdot 28 + 97 = 11297$ days in February, out of which 97 are the 29th. Thus, the actual probability is

$$\frac{97}{11297} \approx 0.00859,$$

slightly less than the initial estimation.

Source: This puzzle was developed by the authors in 2025.

Curiosity. **Friday the 13th**

What is the probability that the 13th of the month is on Friday? It is approximately 1/7. But we can be more precise. The Gregorian calendar has a 400-year cycle. Amazingly, the number of days in the cycle is divisible by 7. We can check that: Non-leap years have $365 = 7 \cdot 52 + 1$ days and leap years have $366 = 7 \cdot 52 + 2$ days. Since there are 97 leap years in the cycle, the total number of days turns out to be $400 \cdot 7 \cdot 52 + 400 + 97 = 7(400 \cdot 52 + 71)$, a multiple of seven! Now, we can just count when the 13th falls on Fridays. This happens exactly 688 times during the cycle. And Friday is the most common day of the week for the 13th. The total number of the 13th is, not surprisingly, 4800. The probability of it falling on a Friday is

$$\frac{688}{4800} \approx 0.14333.$$

You can see that this number is slightly bigger than

$$\frac{1}{7} \approx 0.142857.$$

Source: This curiosity was reported by Julian Havil in his book *Nonplussed!* published by Princeton University Press in 2007.

Solution to Puzzle 8.3 **Can't Beat Nature?**

Solution to question 1:

Even though each family stops having children after a boy, the expected ratio of boys to kids is 1/2, as in any other population. The number of boys in the village will be equal to the number of families; all we need to do in order to find the expected ratio is to calculate the expected number of girls. Half of the families are expected to have no girls, one quarter will be expected to have one girl; continuing this, we find that the expected number of girls in a family is

$$\frac{1}{2}\cdot 0+\frac{1}{2^2}\cdot 1+\frac{1}{2^3}\cdot 2+\cdots=\sum_{n=1}^{\infty}\frac{1}{2^n}\cdot(n-1)=1.$$

This gives the same number of girls as boys on average.

Intuitively, we can understand why that is by thinking about each birth as a new villager instead of a new family member. Each new villager has a 50% chance of being a boy or a girl, regardless of the family from which they are born.

Solution to question 2:

Half of the families in this village have only one kid, who is a boy. Then the average ratio of boys to kids in each family is greater than 1/2. One can calculate this ratio exactly and find that it equals

$$\frac{1}{2}\cdot\frac{1}{1}+\frac{1}{2^2}\cdot\frac{1}{2}+\frac{1}{2^3}\cdot\frac{1}{3}+\cdots=\sum_{n=1}^{\infty}\frac{1}{2^n}\cdot\frac{1}{n}=\ln(2)\approx 0.693.$$

How can it be that the average ratio of boys to kids per family is different from 1/2, the average over the whole village?

Bonus. You are offered a bet: +\$1 if there are more boys in the village, −\$1 if there are more girls, and 0 otherwise. Should you take it?

Source: Folklore.

Curiosity. **Another Strange Village**

Consider another strange village where each family has children until they give birth to exactly one girl more than the number of boys in the family. That is, they can have 1 girl and 0 boys, or 2 girls and 1 boy, etc.

Can you come to terms with the fact that all three claims below are true:

Claim 1. The expected ratio of girls to boys is 1 to 1.

Claim 2. With probability equal to 1, the village will have a finite number of children.

Claim 3. The number of girls is equal to the number of boys plus the number of families.

Source: Developed by the authors in 2025.

Solution to Puzzle 8.4 — This Is Unfair!

Toss the coin twice. If both tosses produce the same result, toss the coin twice again until they are different. We can define HT (heads/tails) as a win and TH (tails/heads) as a loss. Both options are equally probable, thus the probability of winning or losing is 50%.

Bonus. Can you design a game with 1/3 probability of winning, using the same unfair coin?

Source: Different sources seem to credit John von Neumann for this solution and point to the paper 'Various techniques used in connection with random digits,' published in the *National Bureau of Standards Applied Mathematics Series*, 12, 36–38 (1951).

Curiosity. **Faster Resolution**

The algorithm above produces a result as long as $k \geq 0$ exists for which the tosses $2k + 1$ and $2k + 2$ differ in outcomes. A faster algorithm is the following: Assign the sequences HT, HHTT, HHHHTTTT, and so on, to be wins and the reverse sequences TH, TTHH, TTTTHHHH, and so on, to be losses. Start as before, toss two coins: if they are different, the winner is assigned. Toss two more coins: if they are different, the winner is assigned. If not, look at the sequence of all four tosses: if it is one of HHTT or TTHH, the winner is assigned. Continue in a similar manner. This algorithm produces a result as long as the string of tosses is not a constant repetition of heads or of tails. More precisely, if n is the first occurrence of a toss that is different from all the previous tosses, you are guaranteed a result in $2n - 2$ tosses.

Source: Problems of this type have been studied from multiple angles. For the interested reader, we point to one paper that shows many variations "Efficient generation of fair dice with few biased coins" by Luisa Gargano and Ugo Vaccaro, published by *IEEE Transactions on Information Theory* (2002).

Solution to Puzzle 8.5 — Tuesday Glitch

The solution depends on a model.

Solution to question 1:
First, let us recognize that this is a modeling problem disguised as a probability problem. It is a very well-defined modeling problem, but those expecting a fully fledged probability problem might be disappointed and accuse it of being ill-defined or incomplete (more about this later). If we recognize this as a modeling problem, then 1/2, 1/3 or any other result can be plausible so long as there is a plausible model supporting it, and the question of which answer is the most correct is rather a matter of which model is the most appropriate, i.e., the model that relies on the most natural assumptions.

First, we explain where 1/3 came from. If you consider a random family with two children, then there are four equally probable distributions of gender: BB, BG, GB, and GG. Out of those, there are three equally probable options with at least one girl. Only one of them has two girls. Thus, the probability that a random family with two children that has at least one girl has two girls is 1/3. However, why do we assume that the events BG, GB, and GG are equally probable? If the father had children of different genders, he could have said, "I have two kids, one is a boy".

The answer 1/2 is better as it is symmetric to genders. The modeling and counting argument: As a modeling problem, we should count all possible states (defining the sample space as wide as possible), and to define a probability, we need to distinguish all the positive from the negative cases. In this example, a sample is an overheard conversation from a man on the phone.

Men on their phones, in particular fathers, ever so often (as rare as it may be) say "I have x kids, and one is a [boy/girl]".

Modeling Assumption: A natural assumption is that these utterances occur once (or rarely but equally probable) for each son/-daughter a man has, i.e., x siblings will each be referred to this way by their father once (or to the very least with equal probability).

Under this modeling assumption, let us write down every possible utterance of fathers with two kids: There are $2 \cdot 2 = 4$ combinations of boy/girl orders, and there are two kids the father could be referring to. Therefore, we could write down all $4 \cdot 2 = 8$ possible combinations of utterances of fathers with two kids, which can be seen here:

Sample Space		**Order of Kids**			
		BB	**BG**	**GB**	**GG**
Mentioned Kid	First			○	●
	Second		○		●

Now, by counting all positive cases (marked with a black circle) divided by all possible cases (marked with a white or black circle), we obtain $\frac{2}{4} = \frac{1}{2}$.

However, we should repeat that it is possible to create a model to match any answer. Imagine a father who wanted to have a son, and, whenever possible, tries to imply that he has one even if he doesn't. If he had two boys, he would have said, "I have two boys". If he had a boy and a girl, he would have said, "I have two kids, one is a boy". If he had two girls, he would have said, "I have two kids, one is a girl". For this model, the probability of two daughters is 1.

Solution to question 2:
Before we give the answer, notice that we don't know if the first daughter mentioned is the same as the cake thief. If you overlooked this fact, you may want to make a second attempt at solving this riddle.

Our intuition tells us that the more statements the father makes without mentioning any sons, the greater the probability that he doesn't have any.

Again, we can concoct a model producing any answer, but the most natural and unbiased model produces the answer of $\frac{2}{3}$. Let's go straight to the modeling and counting argument:

In this example, a sample is an overheard fragmented conversation from a man on the phone. Now, we are interested in fragmented conversations of the form "I have x kids, and one is a [boy/girl]" followed by "my [son/daughter] stole a cake on Tuesday".

Modeling Assumption: A natural assumption is that these utterances occur once (or rarely but equally likely) for each combination of [son/daughter mentioned first] times [thief son/daughter mentioned second].

Again, we assume that parents of two kids have a distribution of two genders, BB, BG, GB, and GG, with the same probability. In case BB, the probability of the father's statements is zero. In both cases, BG and GB, the probability of the father's statements is 1/4. In case GG, the probability of the father's statement is 1. Overall, the father is twice as likely to have two girls.

Here is the same argument in the table form. Under this modeling assumption, let us write down every possible utterance of fathers with two kids in which one is mentioned in the first part of the truncated conversation, and then one is mentioned in the second part of the truncated conversation. This can be seen in the following table:

		Order of Kids			
Sample Space		**BB**	**BG**	**GB**	**GG**
Mentioned Kid/ Thief Kid	First/First			◦	•
	First/Second				•
	Second/First				•
	Second/Second		◦		•

Now, by counting all positive cases (marked with a black circle) divided by all possible cases (marked with a white or black circle), we obtain $4/6 = 2/3$.

Curiosity. **A Boy Born on a Tuesday**
Gary Foshee gave a very short talk at the 9th Gathering for Gardner in 2010. He said: I have two children. One is a boy born on a Tuesday. What is the probability that I have two boys?

In his mind, the answer was $\frac{13}{27}$. What was his modeling assumption?

Bonus. Note that this result would still be 2/3 if the overheard conversation were "I have two kids, one is a girl" followed by "I have a daughter" since, in this example, it only matters that an additional mention of an ambiguous daughter is made. Even a mention that doesn't add any new information about the daughter is needed. What would happen if multiple mentions of "I have a daughter" are made?

Source: The first question appeared in Martin Gardner's Scientific American column in 1959 with the answer 1/3. He later corrected himself, explaining that the puzzle is ambiguous. Tanya wrote a paper, 'Martin Gardner's Mistake', discussing many modeling strategies for this problem and for the Tuesday-Boy puzzle. The second question is original.

Curiosity. **The Cooperative Principle**
The philosopher of language Paul Grice introduced in 1975 what is known today as the *Cooperative Principle.* The Cooperative Principle is a model describing four *Maxims* of communication expected within *cooperative* conversations. These are:

Maxim of Quantity. *Be as informative as required.*
Maxim of Quality. *Be truthful and faithful to evidence.*
Maxim of Relation. *Be appropriate to the context.*
Maxim of Manner. *Be clear, brief and orderly.*

In the Tuesday Glitch puzzle, we saw how different models could lead to different conclusions. But what if we use a philosophy model, rather than a probability model? From the Maxim of Quantity, we might conclude that "I have two kids, one is a girl" to imply that this person has exactly one daughter. And, independently, when someone says "My daughter stole a cake on Tuesday", the implication is that this person should have exactly one daughter.

We bet this isn't the type of modeling you were expecting to find when you jumped into a probability chapter, was it?

Solution to Puzzle 8.6 Math Department Trips

You may think that the bus drivers are also professors; however, this is not needed to resolve the paradox. If you were one of the professors, when you chose a bus, the rest of the 99 professors were randomly spread between the 10 buses. On average, 99/10 of them will board your bus, which means that your bus, on average, has $1 + 99/10 = 10.9$ passengers. The effect is more significant with only one professor and 100 buses. Then, the average number of passengers in each bus is 1/100, but the professor would always experience one passenger, i.e., themselves.

Source: Gadi Aleksandrowicz's blog (in Hebrew) https://gadial.net/2013/12/13/crowded_bus_paradox/ where credit is given to Ehud Aharoni.

Curiosity. The Hitchhiker

A hitchhiker asked the driver: "How did you agree to give me a ride without knowing me? Aren't you afraid I might be a serial killer?" The driver replied: "It is very unlikely to find two serial killers in one car".

Solution to Puzzle 8.7 — Battleships

Intuition might tell you that since only one of the small ships needs to complete the mission, it is better to send them on the mission. However, it is easy to see that the edge case of $p = 1$ breaks this intuition. If $p = 1$, the two small ships may not complete the mission, but the big ship will complete it for sure. It turns out that it is better to send the big ship for any value of p. The probability that at least one of the small ships succeeds equals 1 minus the probability that they both fail, which equals

$$1 - (1 - p/2)^2 = p - p^2/4.$$

The result is smaller than p, the probability of the big ship's success, for any value of p.

Source: Yogev heard this puzzle many years ago, and the authors could never find a source for it.

Solution to Puzzle 8.8 — Missing 5

The answer depends on Bob's strategy.

Many people assume that Bob loves 5 and is looking for 5 only. Then the expected number of rolls to get 5 or 6 is 3. Only on the rolls without 5 will Bob say something. The answer is 3.

However, there are other natural assumptions. We have two suggestions for Bob's assumptions, where Bob treats every digit that is not 6 equally.

Modeling assumption 1: Suppose Bob lists all the numbers that are missing. Then, when he says that 5 is missing, we are guaranteed that we rolled 1, 2, 3, and 4 before 6. This noticeably increases the number of expected rolls, and the answer is 8.7.

Modeling assumption 2: Suppose Bob randomly chooses a number that is missing. For example, if Alice rolled 1, 2, 3, 2, 1, 6, then Bob notices that 4 and 5 are missing, and mentions 5 with probability 1/2. In this case, the number of expected rolls is 4.26.

Source: This puzzle was credited to Elchanan Mossel in 2017 by Gil Kalai's blog post: https://gilkalai.wordpress.com/2017/09/08/elchanan-mossels-amazing-dice-paradox-answers-to-tyi-30/.

Solution to Puzzle 8.9 — 100 Prisoners, 1 Bullet

Let's simplify the question to *two* prisoners in a field. The prisoners may try to cooperate and run at the same time, thinking each has a 50% chance of escaping. How can we prevent this from happening?

We may tell our least favorite prisoner that if he tries to escape, regardless of what the other prisoner does, we will shoot him. And if the other prisoner tries to run away alone, then we will shoot him instead. The least favorite prisoner surely will not try to escape. The second prisoner knows that, and since he would not try to run alone, he also will not try to escape!

That is how we can prevent two prisoners from escaping, and we can expand this strategy to 100 prisoners. We list them from 1 to 100, and tell them that if some prisoners try escaping together, we will shoot the lowest number. That way, prisoner number 1 will never try to escape, but then prisoner number 2 knows that, so he also will never try to escape, and so on.

Source: The oldest reference we found of this puzzle was posted by user init55 in 2016 in the following forum: https://puzzling.stackexchange.com/questions/27663/discourage-5-prisoners-from-escaping-with-only-1-bullet.

Curiosity. **Tricking the Guard**

After knowing the solution to the 100 prisoners 1 bullet puzzle, we can look for strategies for the prisoners to avoid the guard's strategy and break free. How can they do that? If they are all perfect logicians, they know the strategy that the guard is about to choose. What they don't know is the order of preference that the guard will choose. As long as they don't know the order, the prisoners can try escaping together, and each will estimate his probability of escaping to be 1/100. Therefore, they should all shut their ears while the guard speaks. This might be counterintuitive. In order to escape, they should all avoid consuming extra information provided by the guard. However, each individual prisoner has an incentive to listen to the guard speak because one of them will be his #1 in the list. The prisoners can agree that if one of them is caught listening to the guard, the rest will get rid of him. That way, they are all incentivized not to listen to the guard, and they will all try to escape together.

Curiosity. **Probability of Twins**

The latest data show that in 2022, in the United States, twin births were about 1 in 32 births. An earlier study put the worldwide average at roughly 1 in 84 births. Identical (monozygotic) twins everywhere are born with a fairly constant rate of about 1 in 250 births.

Solution to Puzzle 8.10 **Birthday's Bet**

Here is the standard calculation of the probability that no two people among 22 random people share a birthday. We assume that a year has 365 days and each person's birthday is equally likely and independent.

The first person can have a birthday on any day of the year. Thus, there are 365 choices. The second person must avoid the first person's birthday; thus, there are 364 choices, and so on. The 22nd person has 344 choices. Thus, the probability that they don't share a birthday is

$$\frac{365}{365} \times \frac{364}{365} \times \frac{363}{365} \times \cdots \times \frac{344}{365} \approx 0.524.$$

Tanya seems to have an advantage.

However, there are two important things to consider. First, people are not born randomly; there are seasonal variations, plus there are weekly variations, for example, cesarean sections might not be scheduled on weekends. This will increase the probability of the same birthdays. Second, the group is not random. For example, if a student in the class has a twin, almost surely, the twin would be present. This also increases the probability of the same birthdays.

Suppose the probability of having twins in a random birth is about $\frac{1}{80}$. Let us calculate the probability of no twin births among 22 births. Such a probability equals

$$\left(1 - \frac{1}{80}\right)^{22} = \frac{79}{80}^{22} \approx 0.758.$$

Thus, the probability of having twins in the class is close to 1/4, which significantly outweighs Tanya's advantage in the case of no twins. Tanya's bet is unwise.

If we use the data from the US mentioned in the curiosity above, then the probability of no twins is

$$\left(1 - \frac{1}{30}\right)^{22} = \frac{29}{30}^{22} \approx 0.4743.$$

This number is less than a half!

Source: This puzzle was developed by the authors in 2025.

Curiosity. **Beer Bet**

The puzzle was inspired by an actual event. Many years ago, Tanya was teaching discrete mathematics at Bar-Ilan University (weirdly, in Russian). She had more than 40 students in the class and bet 2 cans of beer that there would be two people with the same birthday. The possibility of twins worked in her favor. After the bet, she collected the birthdays on the blackboard and lost. Looking back, Tanya thinks that it is possible that students wanted Tanya to lose so much that they might have misrepresented their actual birthdays. Who knows?

Chapter 9

Unusual Geometry

A story survives among mathematical circles[1] about the ill fate of a geometer who served the tsar of Russia, known by some as *Ivan the Terrible*. The story goes that Ivan once asked a geometer to estimate how many bricks would be used in the construction of a building that was underway. The geometer was given dimensions and a fair overview of what was to be expected, and with an impressive speed, the mathematically gifted servant of the tsar gave his answer. Not much time passed before the construction was finished and the tsar came back. To his astonishment, the geometer was *exact* in his calculations! Not one more, not one less brick was used.

At this point, the tides took an unexpectedly dark turn. Ivan could not believe that this impressive feat was done by pure mastery of the mathematical arts alone and accused the geometer of sorcery! The man's fate ended then and there as the terrible tsar ordered his execution, freeing Russia from what he believed to be a practicing member of the dark arts.

[1]This story is reported by Malba Tahan in *Matemática Divertida e Curiosa* [11] and is therein attributed to Rebière — *Mathématiques et mathématiciens*, a book from 1889.

Inspired by this story, we invite you to take the playful liberty to go with this idea that geometry contains hidden "dark arts" throughout this chapter. Curiously enough, this is not the only incident when the mathematical sciences have been bundled up with black magic, but perhaps this is something we'll leave for another story... For now, ahead of you awaits the geometry of soda bottles, how to make squares out of circles (kinda), and how to wrap birthday gifts. A rather strange and unusual geometry...

Puzzle 9.1 — A Pentagon in a Cube

Cutting a cube with a single plane can produce a cross-section with the shape of a square. But is it possible to create more complicated shapes?

Question. *Can you cut a cube with a single plane such that the resulting cross-section has the shape of (a) an equilateral triangle, (b) an equilateral pentagon, or (c) an equilateral hexagon?*

Puzzle 9.2 — Largest Regular n-gon

Picture a transparent cube floating in space. You may place tiny drops of paint anywhere on its surface, provided they are *not* all on the same face. Place some number n of paint drops so that they lie at the vertices of a regular n-gon (all sides and angles are equal).

Question. *What is the largest possible value of n for which such a configuration exists?*

Curiosity. — Fermat Point
The Fermat point of a triangle, also called the Torricelli point or Fermat–Torricelli point, is a point such that the sum of the three distances from each of the three vertices of the triangle to the point is the smallest possible.

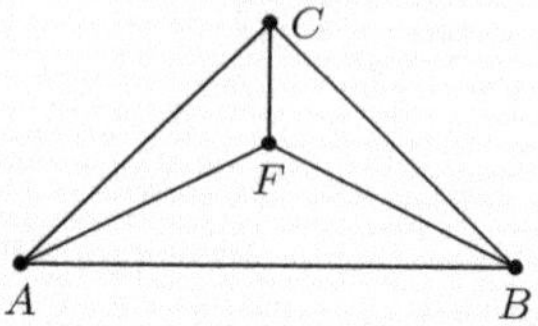

You can imagine the three vertices of the triangle are three villages that you want to connect with the shortest road network. You need to calculate the Fermat point to build these roads. If the triangle has no angles exceeding 120 degrees, then the roads meet at 120-degree angles. In the image above, the Fermat point is F, and the angles between AF, BF, and CF are 120 degrees.

Puzzle 9.3 **The Shortest Road**

A kingdom has four villages located at the corners of a square. The king wants to build a road network that lets travelers go between any two villages. Since paving stones are expensive, he wishes to minimize the total length of road required. The royal adviser proposes paving the two diagonals of the square, forming an X, and claims this is the shortest possible network.

Question. *Is the adviser right?*

Puzzle 9.4 — Second-Cheapest Ribbon

Your friend is celebrating her baby's 2nd birthday, and you've bought a gift that comes in a cube-shaped box. Just like last year, you've decided to decorate the birthday gift with a ribbon. Last year, your ribbon was tied in the cheapest possible way. Since you've recently got a raise, this year you decided to tie the ribbon in the second-cheapest possible way.

Clarification: A decorative ribbon is tightly wrapped around the cube-shaped box in such a way that the ribbon forms a closed loop which cannot be removed without cutting it. The cost of the wrapping is proportional to the ribbon's length (ignore the extra length needed for tying a knot).

Question. *What is the second-cheapest way to tie the ribbon?*

> **Curiosity.** **Second Cheapest**
> People usually don't want to buy the cheapest wine in a restaurant, fearing others might think they are cheap. So, they buy the second-cheapest wine instead. Restaurants know that, so the second-cheapest wine is usually overpriced. Our advice: Either buy the cheapest wine, or a really nice wine.

Puzzle 9.5 — Corners of a Cube

Consider two opposite corners of a unit cube.

Question. *What is the length of the shortest path connecting these corners, going through the faces of the cube?*

Puzzle 9.6 — Two Points on a Cube

The centers of two opposite faces of a cube are connected by four distinct shortest paths, as shown in the picture in different colors and different thicknesses.

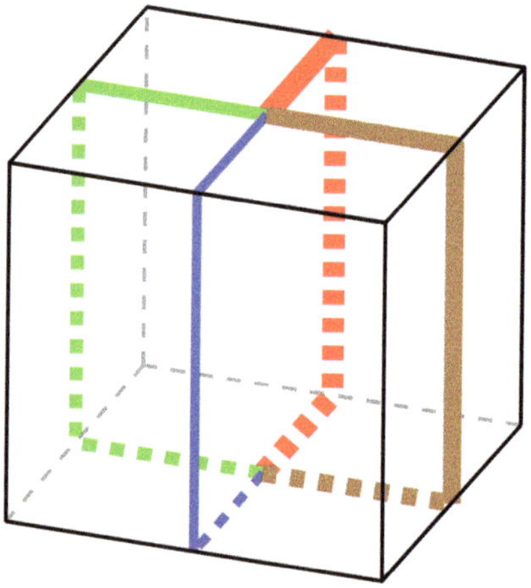

Question. *Can you find two points on the surface of a cube, such that there are exactly three shortest paths connecting them?*

Curiosity. Martin Gardner's Surprise

Glue a cross-shaped piece of paper into a ring with a Möbius strip as depicted in the following image:

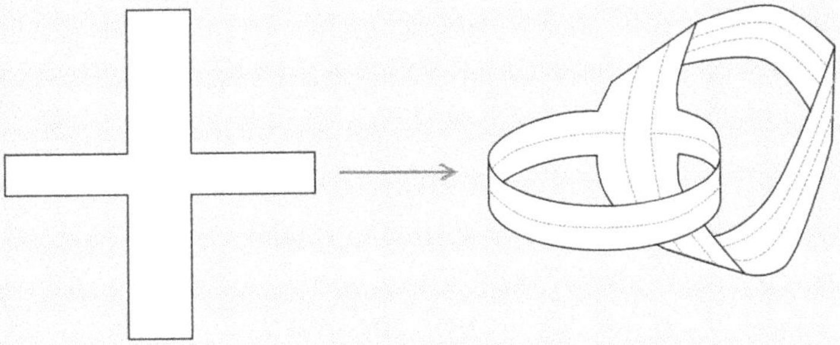

Now, cut the arrangement through the dotted lines as depicted above, trisecting the Möbius strip and cutting the ring almost full way. Flatten the result on the table. What you find might surprise you.

Source: This surprise is an unpublished curiosity by Martin Gardner shared with Ivars Peterson around the year 2000. The occasion and story behind this is shared in his blog, https://mathtourist.blogspot.com/2010/10/martin-gardners-mobius-surprise.html.

Puzzle 9.7 Prince Rupert's Paradox

If you have two cubes of the same size, it is possible to make a square cut through one cube, such that the second cube can go through that hole! In fact, you can make a hole that allows for a bigger cube to go through!

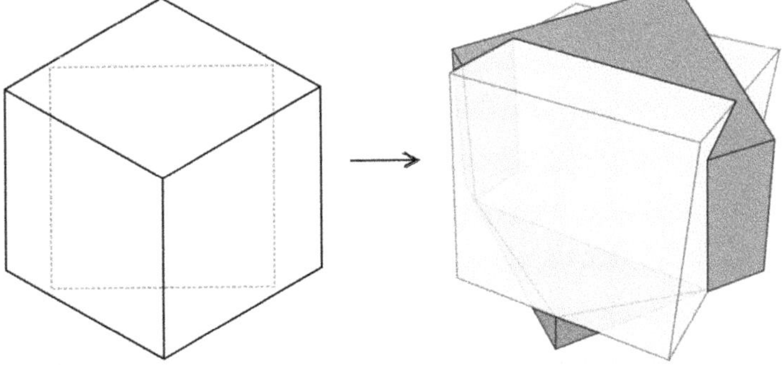

The biggest such cube that can pass through a unit cube is called Prince Rupert's cube, and its side length is approximately 6% larger than the side of the unit cube. It follows that if we take two identical cubes, we can cut holes in both of them, so that they can pass through each other.

Question. *Is it possible to drill a square hole in a sphere and a circular hole in a cube in such a way that the cube can pass through the sphere, and the sphere through the cube?*

Puzzle 9.8 — Soda Bottle Challenge

In some parts of Europe, when you buy a soft drink in a disposable plastic bottle, the lid is tethered to the neck of the bottle in such a way that it does not fall off. Even if you unscrew the cap, it remains physically attached to the bottle's neck by a small plastic strap, so the lid never comes off completely.

Question. *Given two plastic bottles with tethered caps, is it possible to simultaneously screw bottle A to the lid of bottle B, and bottle B to the lid of bottle A without detaching the lids?*

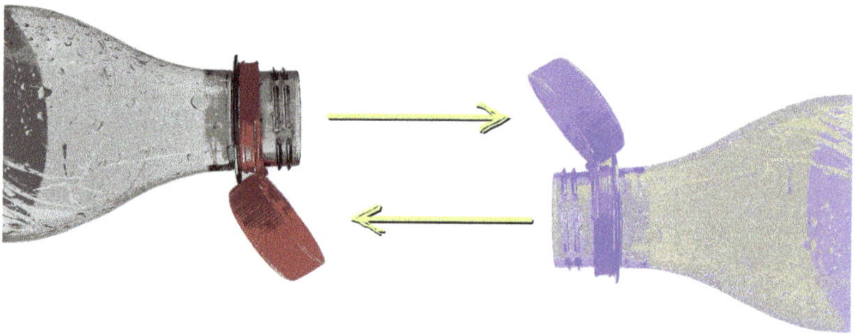

Puzzle 9.9 — The Drunk Wizard

It is said that the first soda bottle was created by a drunk wizard who combined a potion flask with a beer mug.

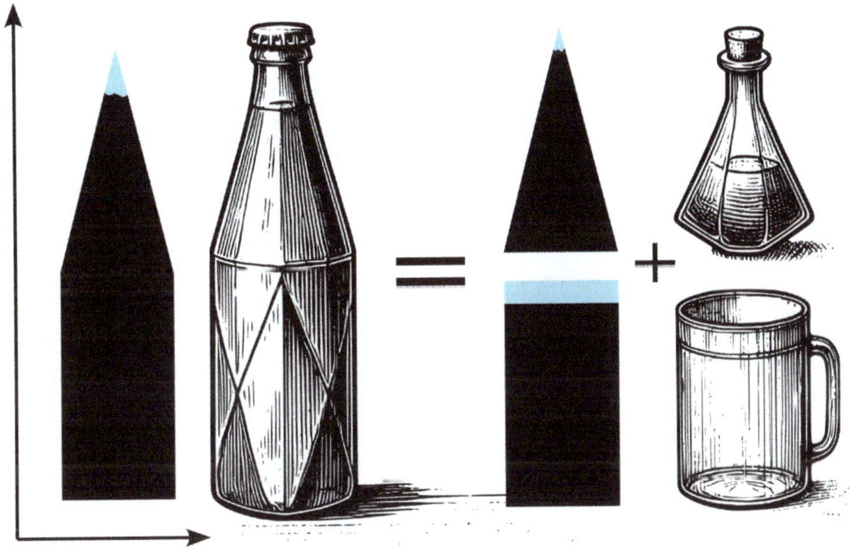

Based on this legend, you are given the following puzzle: Consider a 2D soda bottle, a 2D potion flask, and a 2D beer mug. The mug is rectangular, and the flask is an isosceles triangle; both have width 1 and height 2. The bottle is made by gluing the triangle on top of the rectangle.

Question. *Assuming the three vessels are filled up until 90% of their height, which one must be tilted more until it starts pouring liquid?*

Puzzle 9.10 **One Cut**

Below is a shape made out of nine identical circles. Their centers fall on a square lattice.

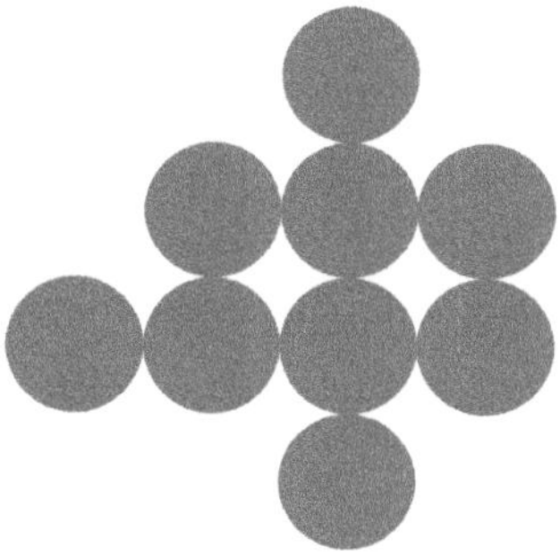

Question. *Find a simple way (without any calculations) to divide the shaded area into two regions of the same area using only a single straight line.*

Solutions

More than two millennia have passed since Euclid in the year 300 BC sat in the ancient city of Alexandria and laid out the rules of the geometric world. Much like an astronomer of his time, his work was concerned with mapping out the constellations hidden in the geometric sciences, uncovering symmetries, and discovering the logic of space. His crown jewel, the legendary mathematical treatise known today as *Euclid's Elements*, is a true grimoire of forms and spatial artistry, and it is therein that the five Euclidean Wizardry Axioms were first set in stone, much like Moses with the ten commandments! This work still echoes strongly in modern geometry, so relevant now as it has ever been, despite the language and ages that separate us from its genesis. However, the last axiom, also known as the Parallel Postulate... we caution you, that behind that one hides a dangerous tale that has led many good wizards to the dark side.

For over a thousand years, geometers have held disputes over the fifth axiom. "Get away with it!" rebellious geometers would say, "Who needs it? It's probably even redundant if we only do the math right", some would say, while others chanted in the background "Unworthy, unelegant, unintuitive!" waving their fists to the sky! And try they did, toiling through lemmas and proofs, however, to no effect. The fifth axiom was never proven to be redundant nor reducible to the other four, despite the many attempts. And then the 1800s arrived, and with it, a new wave of mathematicians.

Euclidean Wizardry Axioms

1. For any two points A and B, there exists a straight line segment connecting them.
2. A straight line segment can be prolonged indefinitely in either direction.
3. For any radius R and center C, there exists a circle centered on C of radius R.
4. All right angles are identical.
5. For any line L and point p not on L, there exists a unique line through p that does not meet L.

It's hard to pinpoint the exact moment that made mathematicians understand that the Parallel Postulate was indeed independent of the rest. But it is certain that the development of working non-Euclidean geometries (geometries that don't follow all the above rules) played a role. And, when geometries that followed rules 1 to 4, but not 5, were discovered, then it eventually became clear that the dark side was wrong all along: Axiom 5 was indeed a worthy, and self-standing rule! It could be bent as any other rule, but it was worthy nonetheless.

In this chapter, in honor of Euclid, we didn't venture outside the confines of standard Euclidean geometry (as some mathematicians in the dark side would have us do). However, we hope you found the rather unusual results we brought to your attention from strange, often-unvisited corners of geometry, amusing!

Solution to Puzzle 9.1 **A Pentagon in a Cube**

We can get a square by cutting parallel to one of the faces. Another way to cut is perpendicular to the main diagonal. If you cut close to one of the vertices, you will get an equilateral triangle.

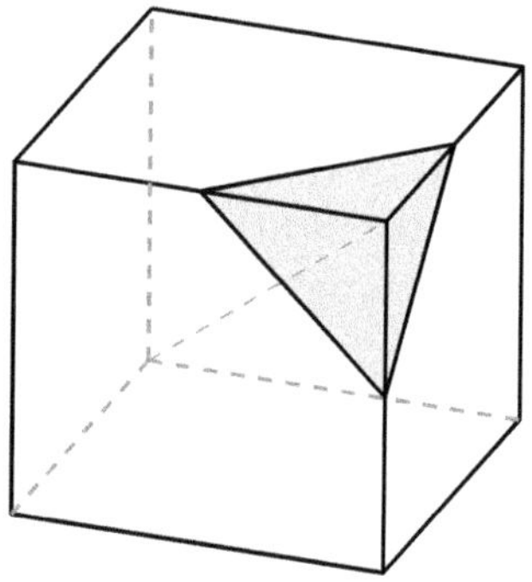

If you cut through the center of the cube, you will get a regular hexagon.

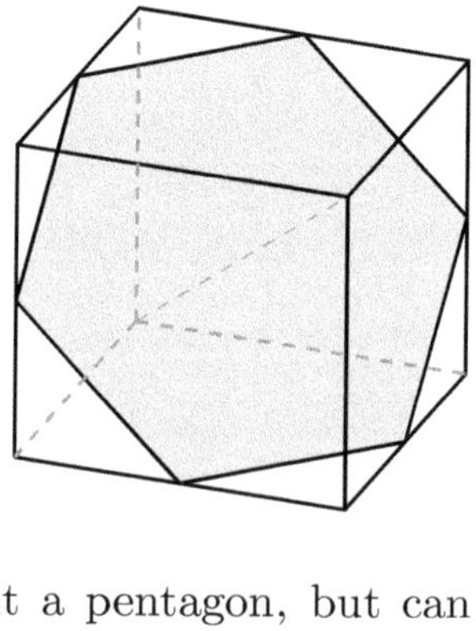

It is possible to cut a pentagon, but can we cut an equilateral pentagon? The cube has three pairs of parallel faces. That means if we cut a polygon with more than three sides, at least one pair of the sides should be from the opposite faces. It follows that those sides are parallel. Thus, if we ever cut a pentagon, it has to have two pairs of parallel sides: we can never have a regular pentagon.

Source: Folklore.

Solution to Puzzle 9.2 — Largest Regular n-gon

We can't have more than two points on the same side, because if we have three non-collinear points, they define a plane, and our polygon will belong to one face. It follows that the polygon has no more than 12 vertices: two for each of the six faces.

Now, we construct a polygon with 12 vertices. We showed in the previous problem how to cut the cube to get a regular hexagon. Now we just need to find 12 points on a regular hexagon that form a regular 12-gon. This is easy to do.

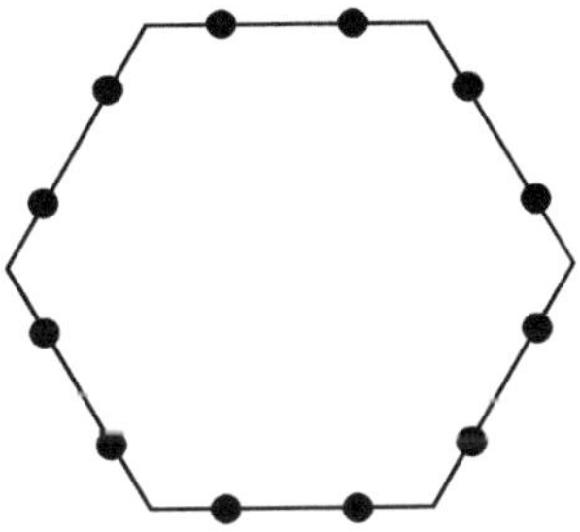

Source: Folklore.

Solution to Puzzle 9.3 — The Shortest Road

Suppose the vertices of the square are A, B, C, D, and the center is M. Then triangle ABM is right. That means we can find the Fermat point inside the triangle, and the new roads connecting vertices A, B, and M will be shorter than the current connection. The answer is no: this set of roads is not the shortest. The shortest roads have to meet at angles of 120 degrees, as in the picture.

If the side of the square is 1, then the length of the two diagonals is approximately 2.828; the minimum length of the road is 2.732.

Source: Folklore.

Curiosity. **Road for a Pentagon**

Here is the minimal road network for a pentagon.

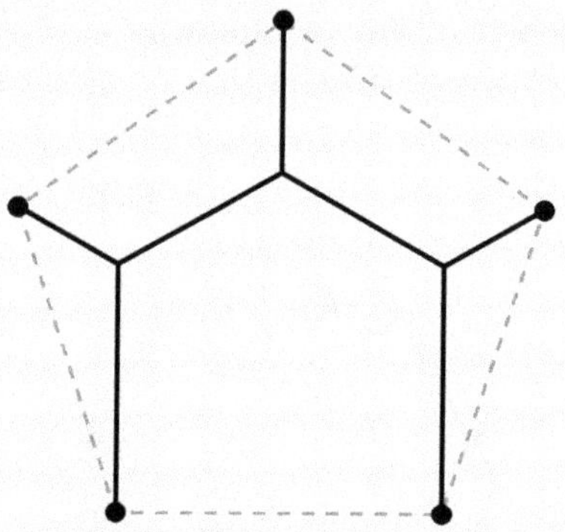

Solution to Puzzle 9.4 **Second-Cheapest Ribbon**

Label three adjacent faces of the cube as A, B, and C, and let A', B', and C' be their opposite faces.

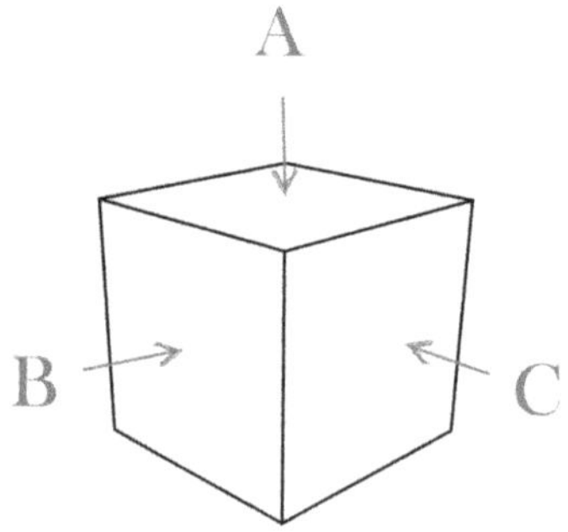

Note that the ribbon must go through all the faces, or else it will slide off. Let us consider all the possible sequences of faces being visited by the ribbon's path (e.g., $ABA'CAC'B'$). Sequences follow specific rules:

- Consecutive repetitions of faces are not allowed: for example, AA is not allowed as it is the same as A.
- A face can't be followed or preceded by its opposite: for example, AA' is not allowed as we need to pass other faces to get to A' from A.

In order to be as sparing as possible of ribbon costs, it is natural to restrict our search to sequences in which the ribbon visits each face only once.

Now, we only look at letters, ignoring the apostrophe. We show that three consecutive different letters can always be found. Suppose letter C is the last to appear in our string. Consider the first appearance of C. That means we have both letters A and B appear right before C. By the rules above, the two letters right before C have to be different. Thus, we found a sequence of three consecutive letters that are different.

As the sequence of faces $F_1F_2F_3F_4F_5F_6$ is equivalent to the sequence $F_2F_3F_4F_5F_6F_1$, we can start our string at any place, so we start with three different letters. Thus, without loss of generality, we can assume that the string starts with ABC.

The next letter after ABC could be A' or B'. Thus, we have options $ABCA'B'C'$, $ABCA'C'B'$, $ABCB'A'C'$, and $ABCB'C'A'$. The last is impossible as A and A' are next to each other. If we start the third string from the third letter, we get $CB'A'C'AB$, which, upon renaming the faces, is equivalent to the second string. Thus, we find that these are all possible sequences:

1. $ABCA'B'C'$.
2. $ABCA'C'B'$.

Type 1 produces a hexagonal path around the cube, as illustrated below, looking from two different angles.

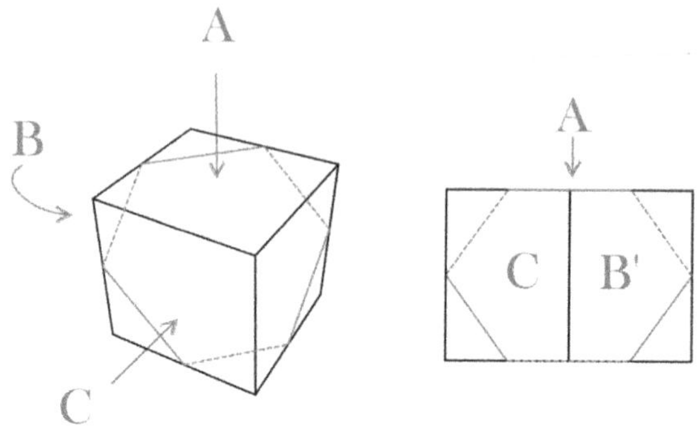

Type 2 produces a distorted (not in the same plane) hexagonal path around the cube as illustrated as follows, looking from the same two angles.

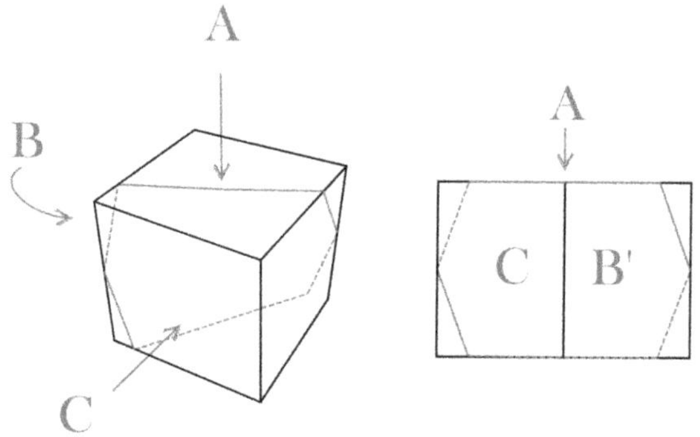

The lengths of these paths are perhaps easier to calculate with the help of the auxiliary midlines. We illustrate this with the figure below; to the left, we show Type 1, and to the right, Type 2:

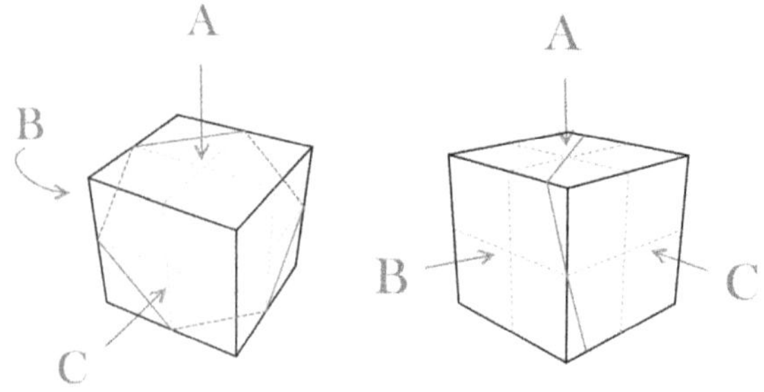

Each line segment in Type 1 is $\frac{\sqrt{2}}{2}$. Thus, the length is $3\sqrt{2} \approx 4.243$, which is the shortest length. For Type 2, the segment passing through side B and C is $\sqrt{\frac{1}{4}+\frac{1}{16}} = \frac{\sqrt{5}}{4}$. The length of the segment passing through side A is twice that. The total length is $2\sqrt{5} \approx 4.472$, which is the second shortest.

Source: This was developed independently by the authors and the authors later discovered that the same problem was tackled in the paper: 'Closed geodesics on regular polyhedra' by Dmitrii Fuchs and Ekaterina Fuchs, published in *Moscow Mathematical Journal*, 7.2 (2007): 265–279.

Solution to Puzzle 9.5 — Corners of a Cube

Imagine that the cube is made out of paper. When we unfold it, the shortest distance is the diagonal of two neighboring faces of the cube. The distance is thus the square root of 5.

Source: Folklore.

Bonus. Consider a general cuboid with dimensions $A \times B \times C$. Show that the most distant point from one corner is not always the opposite corner. In particular, the question of finding the most distant point on a $1 \times 1 \times 2$ cuboid is known as Kotani's ant problem and was discussed by Martin Gardner in 'The Ant on a $1 \times 1 \times 2$', *Math Horizons*, February 1996, 8–9.

Solution to Puzzle 9.6 — Two Points on a Cube

Assume for simplicity that the cube is a unit cube. Unfold the cube and mark the centers of the faces (in our following illustration, we mark them with a "+"). Now, pick one face and draw a circle of diameter 2 centered on one of the marked points. By construction, this circle goes through the center of the four adjacent faces. Pick two opposite points on the circle close to the centers of faces, but not the center of the faces. Two such points are illustrated below with gray dots.

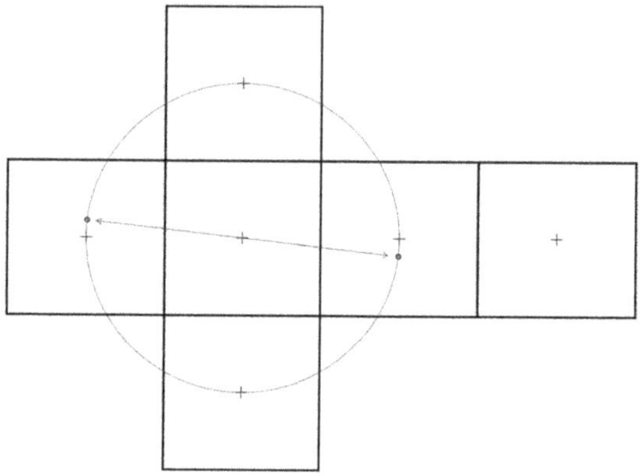

These two points have a minimum distance of 2 going through the center of the circle and also going through the adjacent faces, as illustrated in the following figure.

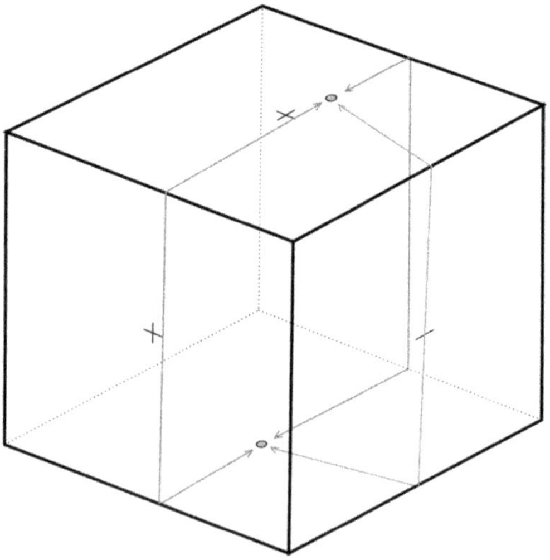

These are the only three minimum distance paths connecting these points since the path going through the face opposite from the center of the circle is more than 2 units long.

Source: This puzzle was developed by the authors in 2024.

Bonus. Does there exist a pair of points such that there are exactly five different paths with minimum distance?

Solution to Puzzle 9.7 — Prince Rupert's Paradox

Unfortunately, the answer is no!

The largest square you can drill through a sphere of diameter D has side L that must satisfy the equation $\sqrt{2}L < D$. This is illustrated below.

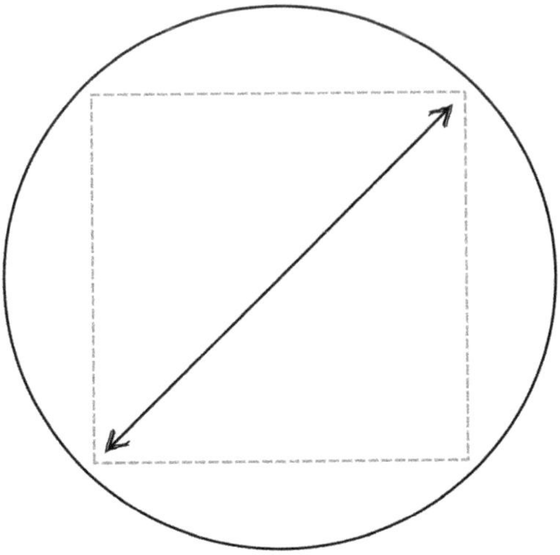

And the largest circle you can drill through a cube is obtained by drilling from one vertex to its opposite, as illustrated below.

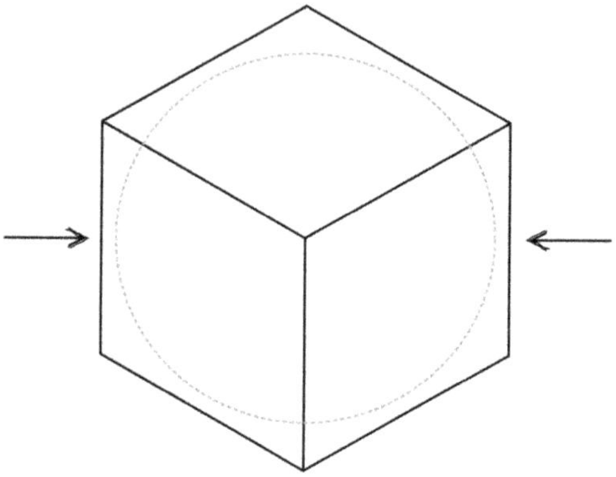

In this view, the shadow of the cube forms a hexagon. In this projection, the big diagonal of the rhombus keeps its length and is equal to $\sqrt{2}L$. That means the side of the hexagon is $\sqrt{\frac{2}{3}}L$. The radius of the inscribed circle equals the height of an equilateral triangle whose side is the same as the side of the hexagon. Thus, the radius is $\frac{\sqrt{3}}{2}\sqrt{\frac{2}{3}}L = \frac{L}{\sqrt{2}}$.

Now, note that in this view, the distance between two opposite edges limits the maximum circle diameter, as illustrated by the arrows. The diameter must obey: $D < \sqrt{2}L$. This condition contradicts the condition we found at the beginning.

Source: This is an original puzzle inspired by the cube curiosity named after Prince Rupert of the Rhine.

Curiosity. **Lonely Solids**

In the *Prince Rupert's Paradox* puzzle, we challenged you to verify if a cube and a sphere can "Rupert" through each other. And, as shown above, the answer is no! However, much more can be said:

We say that a 3D solid is *lonely* if no other solid can "Rupert" through it. Here is a proof that the sphere is, in fact, a *lonely* solid.
Assume by contradiction that there exists some solid S that could "Rupert" through a sphere of diameter D. Because that solid can go through the sphere, it must therefore be contained within a cylinder of diameter $D' < D$ and infinite height. And, since the sphere can go through S, it must also be able to go through this cylinder of diameter D'. However, an infinite cylinder of diameter D', no matter which angle you approach it with, will never allow for a circular hole of diameter D without breaking it apart. Contradiction!

Bonus. Can you come up with another lonely solid?

Solution to Puzzle 9.8 — Soda Bottle Challenge

Surprisingly, this can be done! First, screw the lid of the first bottle to the second bottle all the way in. Then, place the lid of the second bottle on the opening of the first one, and rotate the bottles around each other.

Here is a little party trick for you: challenge your friends to solve this puzzle. After they solve it, ask them if they can solve it with their hands glued to the bottles. The secret? Align the caps and bottles, and in a pedaling motion with your arms, spin the bottles while keeping the caps in contact with the opposite bottles. Surprisingly, they will attach just as before!

Source: This puzzle was developed by the authors in 2024.

Solution to Puzzle 9.9 **The Drunk Wizard**

The answer is the soda bottle! To prove this, first, notice that by the time the beer mug is half empty, neither the bottle nor the flask will have even started to pour their contents, as seen in the following image.

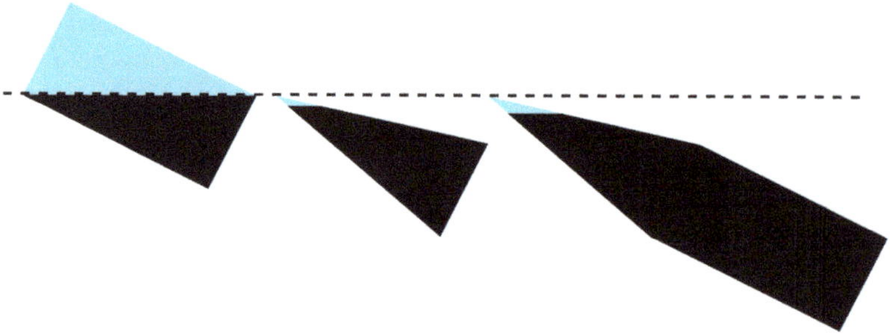

The question remains as to whether the bottle or the flask pours the top 10% of their contents first. The trick is to rescale the flask to be of the same height as the bottle. Notice that the shape at the top 10% is the same for the soda bottle and the rescaled flask. This way, *the top 10% of the height corresponds to the same area for both the bottle and the rescaled flask.*

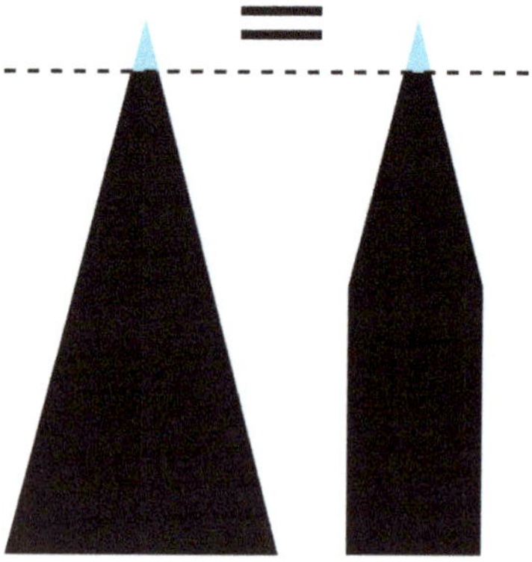

Now, consider the area above the tip of the tilted vessels as in the following illustration.

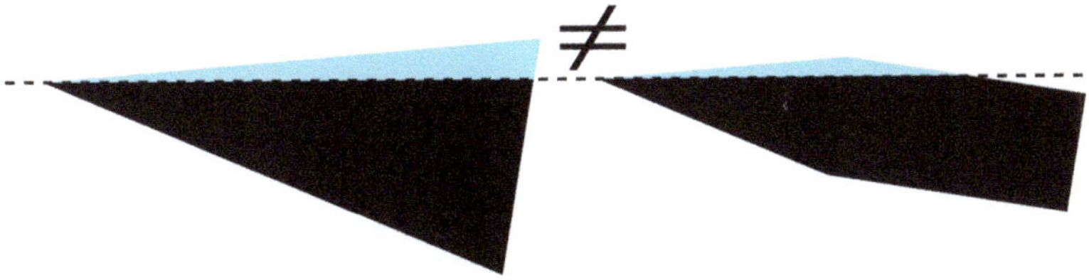

Clearly, at any given angle, the flask will have poured more water than the bottle. Therefore, the flask starts pouring liquid before the bottle.

Source: This puzzle was developed by the authors in 2024.

Bonus. Under the same conditions, do the 3D vessels require more tilting or less tilting to pour liquid?

Solution to Puzzle 9.10 **One Cut**

One trick to divide the area is to first add an imaginary circle that restores the symmetry of the shape:

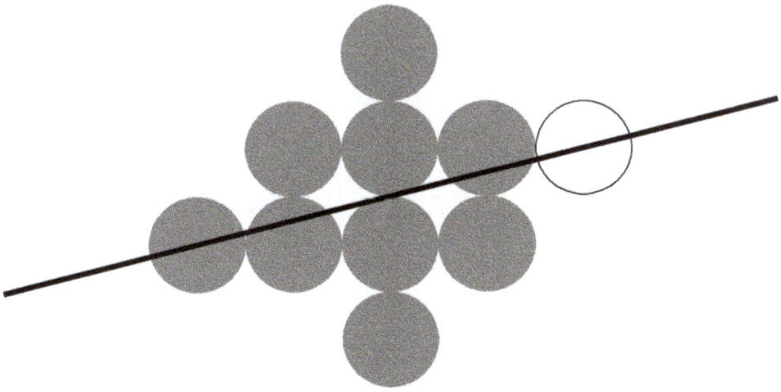

The resulting figure has rotational symmetry. Thus, any line passing through the center of symmetry will divide it into two regions of the same area. For our purposes, we need a line that divides the new circle in half, so that the extra circle contributes the same area to both regions. Therefore, our line should be the line that passes through the center of the extra circle and the center of symmetry.

During testing this puzzle, Dan Klain suggested a different solution presented in the picture:

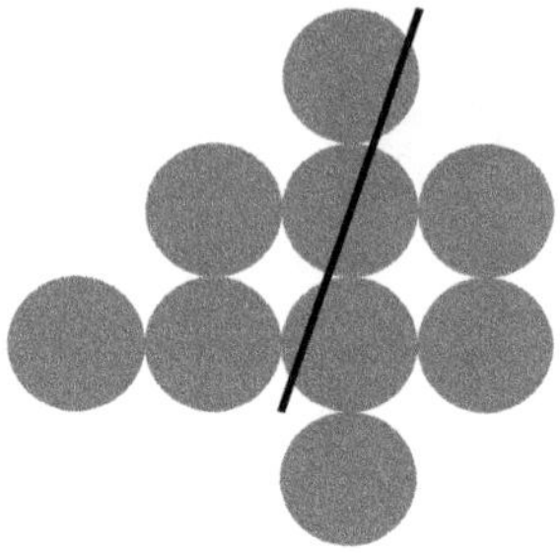

Source: This puzzle was developed by the authors in 2024.

Bonus. It seems like every brainteaser book should have at least one toothpick-moving puzzle. We would like to end the book with one of our favorite puzzles by Martin Gardner. It looks so simple, yet the solution is quite elusive.

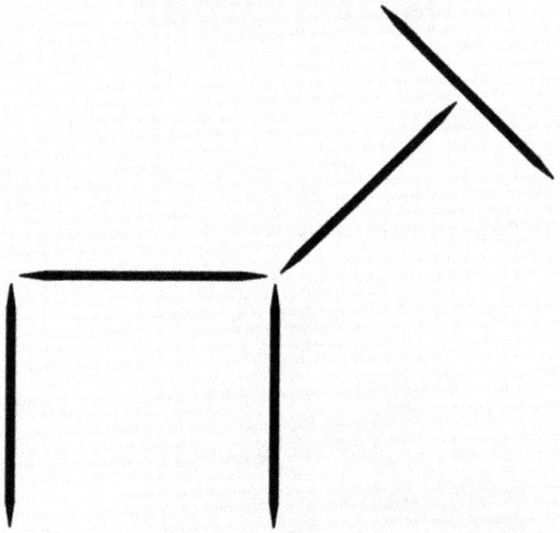

In the figure above, there is a giraffe made out of five identical toothpicks. Can you move one toothpick to obtain the same giraffe from a different angle?

Acknowledgments

David is grateful to his wife, Anna-Marie, for countless moments of support that have allowed him to chase down mathematical rabbit holes, many of which eventually became puzzles in this book. He is grateful to Hugo Rosado and Saar Zehavi for many hours of helpful discussions on dice projects and related mathematical ideas. The *Pythagoras Club* from college times and his dad were greatly influential in his love for puzzles and the exploration of strange mathematical variations. Finally, professors Takahashi, Ernesto, and Adil have played important roles in David's passion for math, science, and creative problem-solving.

Tanya is grateful to her sons, Alexey Radul and Sergei Bernstein, for lively puzzle discussions and for tackling the hardest challenges. She gives special thanks to Alexey for his generous help with English, often while patiently explaining its rules. Tanya is indebted to her grandchildren, Lev and Alex, and to her PRIMES STEP students for test-solving many of the puzzles and inspiring her to work harder. She also thanks Dan Klain for insightful conversations about geometry puzzles.

Yogev is grateful to his wife, Yael, for the many hours spent discussing and testing new puzzle ideas. He also thanks his baby girl, Avigail, whose bedtime routines kept him awake long enough to dream up new puzzles. Finally, he is grateful to his puzzler friends whose shared puzzles over the years served as inspiration for many of those collected here.

All three authors acknowledge the help and support of the World Scientific team, starting with Rochelle Kronzek Miller, Srinidhi Murugan, and Nambirajan Karuppaiah, editors in our project. In addition, we thank the many unnamed World Scientific team members who also silently made this possible.

Bibliography

[1] OEIS Foundation Inc. The on-line encyclopedia of integer sequences, 2023, published electronically at https://oeis.org.

[2] F. Bartoš, *et al.* Fair coins tend to land on the same side they started: Evidence from 350,757 flips. *Journal of the American Statistical Association*, 1–10, 2025.

[3] G. Gamow. *One Two Three... Infinity: Facts and Speculations of Science*, New York: Viking Press, 1947.

[4] R. Smullyan. *What Is the Name of This Book?*, New York: Touchstone Books, 1986.

[5] E. W. Weisstein. *Almost Integer.* https://mathworld.wolfram.com/AlmostInteger.html, 2002.

[6] R. P. Savage, The paradox of nontransitive dice. *The American Mathematical Monthly*, 101(5), 429–436, 1994. https://doi.org/10.2307/2974903.

[7] S. Eldridge, I. D. de Oliveira, and Y. Shpilman, *The Paradox of Anti-Inductive Dice*, 2025. Preprint at https://arxiv.org/abs/2503.16306.

[8] R. Ford, J. Grime, E. Harshbarger, and B. Pollock, Go first dice for five players and beyond. *Recreational Mathematics Magazine*, 10, 75–87, 2023. https://doi.org/10.2478/rmm-2023-0004.

[9] M. Gardner, *The Incredible Dr. Matrix*, New York: Charles Scribner's Sons, 1976.

[10] M. Tahan, *The Man Who Counted*, W. W. Norton & Company; Reissue edition (May 2015), translated by Alastair Reid and Patricia Reid Baquero from *O Homem que Calculava*, A.B.C. do Rio de Janeiro 1938.

[11] M. Tahan, *Matemática Divertida e Curiosa*, São Paulo, Brazil: Editora Record, 2014.

[12] P. Schultheiss, S. S. Nooten, R. Wang, M. K. L. Wong, F. Brassard, & B. Guénard, The abundance, biomass, and distribution of ants on Earth. *Proceedings of the National Academy of Sciences of the United States of America*, 119 (40), e2201550119, 2022. https://doi.org/10.1073/pnas.2201550119.

[13] H. Nover and A. Hájek, Vexing expectations. *Mind*, 113 (450), 237–249, 2004.

Copyright Statements of Images and Glyphs

Ornaments in between chapters + Puzzles 4.1 to 4.9: Copyright 2016—2020 Alain Matthes — *The LATEXProject Public License 1.3.*
Separator in Puzzle 1.1: Copyright Twemoji symbols © Twitter, Inc. — *CC-BY 4.0* license.
Separator in Puzzle 1.2: Copyright of the "Ticket" icon from Font Awesome — *CC Attribution 4.0 International* license.
Separator in Puzzle 1.3 & 1.6: Copyright Noto Emoji symbols © Google LLC — *SIL OPEN FONT LICENSE Version 1.1.*
Separator in Puzzle 1.4: Copyright Noto Color Emoji symbols © Google LLC — *SIL OPEN FONT LICENSE Version 1.1 — 26 February 2007.*
Separator in Puzzle 1.5: Copyright EmojiOne Color 2016 © Adobe Systems Incorporated — *MIT License.*
Separators in Puzzles 2.1 to 2.11: Copyright Abbild Font by Vladmir Nikolic — *Free for commercial use* license by 1001fonts.com.
Separators in Puzzles 2.1 to 2.11: Copyright Dancers CTAN font package by Alan Stanier — *Free for commercial use* license by CTAN.
Separators in Puzzles 3.1 to 3.14: Copyright Semaphor font by Vit ZYKA — *SIL Open Font License.*
Separators in Puzzles 4.1 to 4.9: Copyright Custom Dice CTAN Package by Peter Rowlett — *CC-BY SA 4.0* License.
Separators in Puzzles 5.1 to 5.6: Copyright Segoe UI Emoji font © Microsoft Corporation — Used under the Windows font embedding terms.
Separators in Puzzles 6.1 to 6.7: Copyright Briaroak Shire font by Edward Buri in collaboration with Unauthorized Type — *Free for commercial use* license by 1001fonts.com.
Separators in Puzzles 7.1 to 7.5: Samys Keys'N'Keys font by RetSamys — Public Domain.
Separators in Puzzles 8.1 to 8.10: Copyright Texticles font © Graham Meade/Apostrophic Labs — *Free for commercial use* license by 1001fonts.com.
Separators in Puzzles 9.1 to 9.10: Tetris Blocks 2.0 font by Carl Heavenezer M. Castro — Dedicated to the Public Domain.
Pinocchio image in Chapter 1: Copyright Archivio storico della Libreria editrice Paci LA TIFERNATE e della Casa editrice IL SOLCO a.p.s. Creative Commons Attribution Share Alike 3.0 Unported.
Tupper's self-referential formula in Chapter 1: Copyright Wikipedia User Larske — Creative Commons Attribution-Share Alike 3.0 Unported license.
Broken calculator numbers in Chapter 3: by User Firkin in OpenClipArt.org — Public Domain.
Numbers 19 and 7 in Chapter 3: from "Blomstersiffror med rim" (1894) by Ottilia Adelborg — Public Domain.
Opening image in Chapter 4: from "aer perspectiva corporum regularium", a manuscript from the 15th century in the Public Domain.
Image of Owl on a solid in Chapter 4: from "Geometria", a manuscript from 1543 illustrated by Augustin Hirschvogel, Public Domain.
Dance notation in Chapter 7: from the manuscript "Choregraphie" illustrated by Raoul-Auger Feuillet in 1701, Public Domain.
Giraffe matches in Chapter 9: Built from LUSIVERS Font copyright CROL:rene.NL — *Free for commercial use* by online-fonts.com.
Euclid font: Magic School One Font © 2004 Michael Hagemann www.FontMesa.com — *Free commercial license* by dafont.com.
Euclid image: by Andre Thevet 1584 — Public Domain.
Hungarian bank note image: edited from widely available images.
All other images were self-generated and all rights are reserved by the authors.

About the Authors

Ivo Fagundes David de Oliveira is a mathematician, inventor, and educator with a passion for puzzles, games, and riddles. With a master's degree in Mathematics and a Ph.D. from the University of Minas Gerais, he has spent over a decade teaching and inspiring students of all ages. David is known for his engaging presentations and his invention of paradoxical objects, including his "Prophet Coins" and the "Triple Gozinta Boxes" (released by TCC Magic in cooperation with DistantLands.Studio). His inventions celebrate the beauty and play found in unexpected corners of mathematics.

Tanya Khovanova is a renowned mathematician and blogger. She holds a Ph.D. in Mathematics from Moscow State University and currently works as a Lecturer at MIT. Tanya conducts research in recreational mathematics and works with several programs that help youngsters do research in mathematics. Tanya manages her personal website at tanyakhovanova.com, her math blog at blog.tanyakhovanova.com, and a website with a huge collection of number properties at numbergossip.com.

Yogev Shpilman holds a Ph.D. in Particle Physics from the Weizmann Institute of Science. He has been collecting puzzles since childhood and has a lasting interest in number theory, logic, and games, such as chess and Go. He is currently developing a puzzle-focused video game inspired by these interests.

Index

www.ingramcontent.com/pod-product-compliance
Ingram Content Group UK Ltd.
Pitfield, Milton Keynes, MK11 3LW, UK
UKHW020311150926
8624IPUK00014B/513

9 789819 822362